T0074767

BENNU 3-D

ANATOMY of an ASTEROID

BENNU 3-D
ANATOMY of an ASTEROID

DANTE LAURETTA BRIAN MAY

CARINA A. BENNETT, KENNETH S. COLES, CLAUDIA MANZONI, C.W.V. WOLNER

THE
London Stereoscopic Company, LTD.

Published in 2023 by
The London Stereoscopic Company (LSC)

A catalogue record for this book is available
from the British Library.
ISBN 978-1-8381645-7-7

Published in North America by
The University of Arizona Press (UAP)
Main Library Building, 5th Floor
1510 E. University Blvd, Tucson, AZ 85721-0055
www.uapress.arizona.edu

Library of Congress Control Number: 2023933139
ISBN 978-0-8165-5176-7

The London Stereoscopic Company
Director – Brian May
Publisher and Editor-in-Chief – Robin Rees
Art Director – James Symonds
Cover Design – Brian May
Stereoscopy – Brian May and Claudia Manzoni
LSC Manager – Sara Bricusse
PR Manager – Nicole Ettinger
Archivist – Denis Pellerin
Website – Phil Murray
Proofreader – Cathy Lowne

OSIRIS-REx
Graphics Specialist – Heather L. Roper

Book printed and bound in China by
Jade Productions.
OWL Stereoscope designed by Brian May.
First printing 2023.

www.LondonStereo.com
BrianMayForReal
@LondonStereo
@LondonStereo
@LondonStereo

Disclaimer
Every effort has been made to correctly credit the images in this work. If there are any wrong attributions, please contact the publishers.

On its way to Bennu, OSIRIS-REx captured this image of its sample collector in the foreground and Mars as a tiny point of light in the distance.

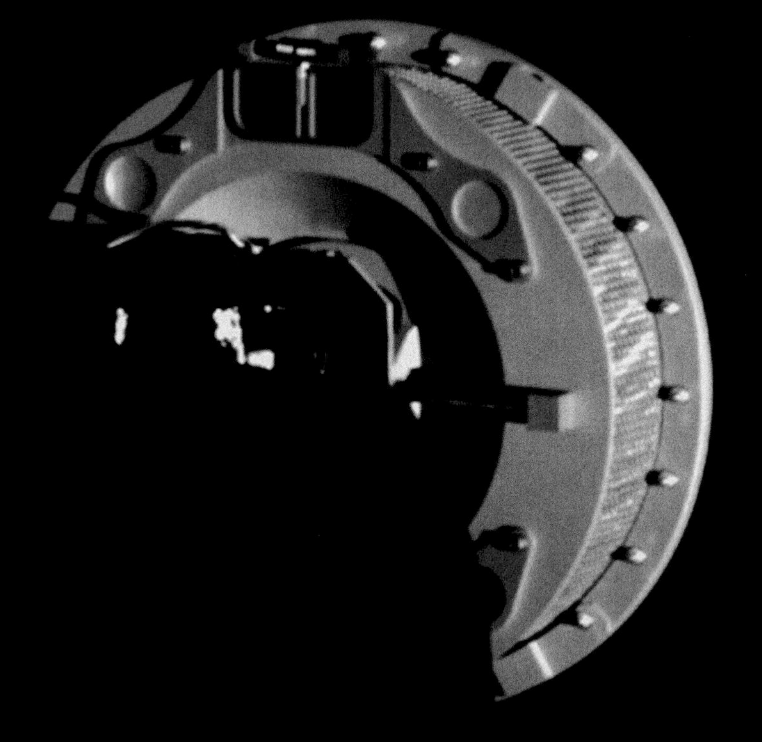

Contents

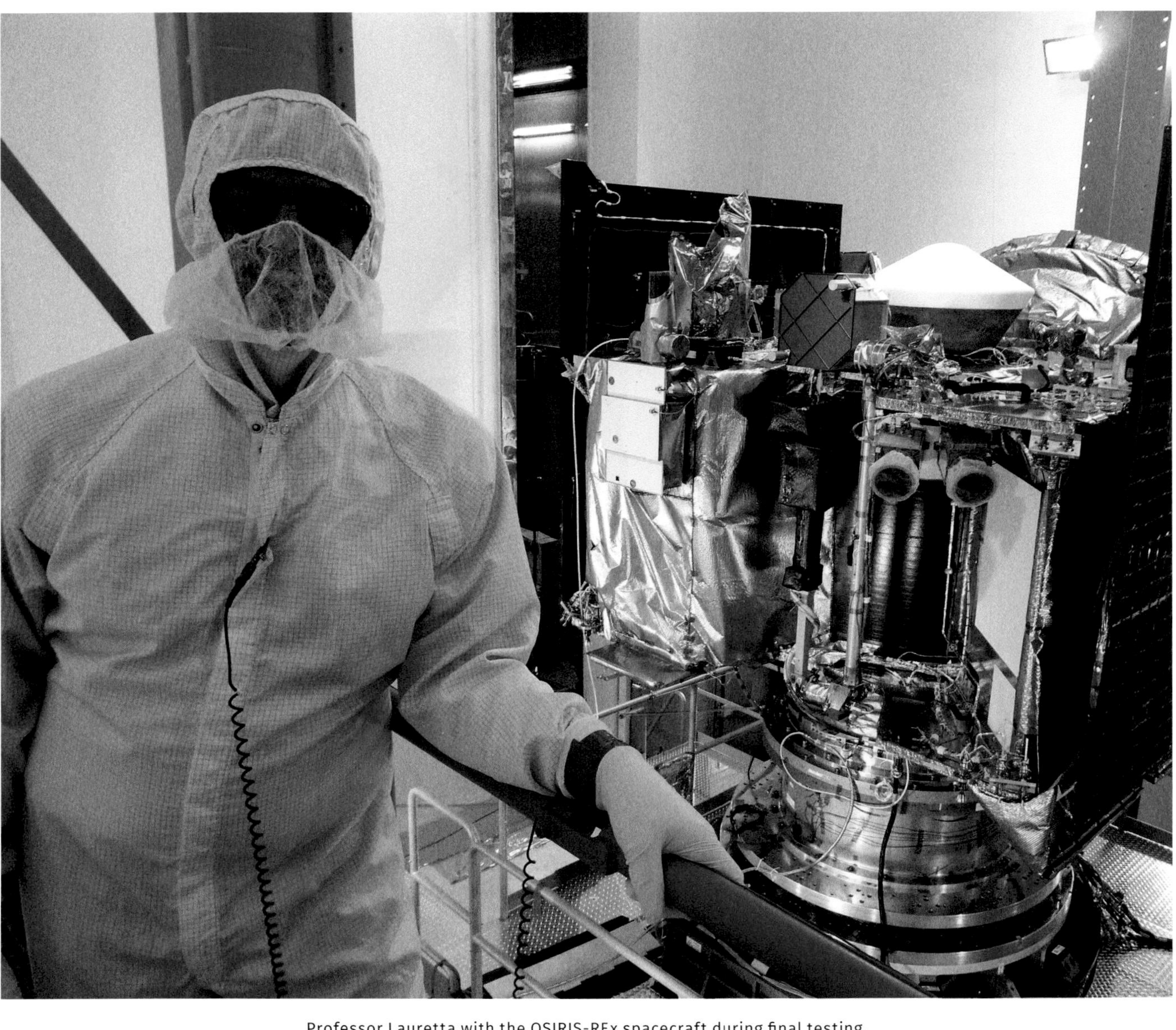

Professor Lauretta with the OSIRIS-REx spacecraft during final testing.

Preface by Dante Lauretta

I was first introduced to Dr. Brian May in April 2016 by our mutual friend Grigorij Richters, co-founder of Asteroid Day. Brian and I were both early supporters of this movement, which is a United Nations-sanctioned global awareness campaign to inspire, engage, and educate the public about asteroids' opportunities and risks. Brian and I corresponded briefly about the mission and my hometown of Tucson, Arizona, where he had spent some time enjoying the natural beauty of the Sonoran desert and using it for self-reflection, as many do.

I was, of course, a huge fan of Brian's music. The early 1980s were a tough time in my life, and I found respite in music. The song *Under Pressure* helped me through some really difficult situations. The fact that I was corresponding with one of my childhood heroes was beyond cool. However, OSIRIS-REx is serious business, and I was not interested in tourists joining the team simply because they were celebrities. Our correspondence went quiet for about a year, and I figured that Brian had other priorities, and I could at least tell my kids that I had corresponded with a rock star.

No sooner had I written off this communication as fantasy when I received an email inviting me backstage to a Queen + Adam Lambert concert in Phoenix in June 2017. I quickly grasped the opportunity and took my family to enjoy the show. Besides meeting Brian backstage (the first time we met in person), the highlight of the show for me was Brian's guitar solo, with images of the Rosetta spacecraft exploring its comet in the background. Space and rock & roll — the perfect combination.

After that unforgettable night, Brian and I stayed in touch, exchanging occasional messages about space and music. As the OSIRIS-REx mission progressed, I couldn't help but share some of the latest developments with him. To my delight, Brian showed a keen interest in the mission and the science behind it. It was clear that he was not just a casual fan, but a true space enthusiast and an advocate for space exploration.

OSIRIS-REx is a NASA mission led by the University of Arizona to study Bennu, a carbonaceous asteroid with a diameter of about 500 meters, and return a sample to Earth for analysis. The mission has five primary scientific objectives that make up the name: Origins (where did the building blocks of life come from), Spectral Interpretation (how well do telescopic data inform us about asteroid properties), Resource Identification (do near-Earth asteroids contain economically valuable material), Security (what risk do asteroids like Bennu pose to humanity), and Regolith Explorer (what can we learn from directly probing Bennu's surface layer of rocks and dust). Each of these topics is crucial to understanding our neighbors in the Solar System and the potential for life beyond Earth.

On November 18, 2018, the first resolved images of Bennu started coming down from the spacecraft. I was amazed by the results that Brian and his collaborator Claudia Manzoni produced by processing our data into stereo images, allowing us to see Bennu's rugged and rough landscape in glorious 3-D.

Seeing Bennu's surface in this way really brought home the intimidating reality of this asteroid. It was far beyond our initial spacecraft design capabilities. At first it seemed like our task was impossible, that we were never going to find a suitable location to collect our sample. However, it was clear that stereo imaging was a valuable tool for our mission, and I knew that I wanted to put Brian to work.

I offered Brian and Claudia formal positions on our science team, providing them with full access to our data, beyond what we were ready to put out in public at that time. After onboarding him to the team, he wrote to me and said:

"Dante, this really means a lot to me. I'm used to being on the edge of things like this. To be accepted as a collaborating scientist rather than a celebrity would suit me much better. I'm a worker. I get my fulfilment from creating great things — or being part of a team that creates." "Well," I thought to myself, "let's see if he means it." And it turned out he did.

Brian and Claudia worked tirelessly to process countless stereo images of regions on Bennu. They were committed to our mission, regardless of whether Brian was on tour, performing at the Oscars, or relaxing on a beach somewhere. Their hard work ultimately helped us to reveal a critical fact about Bennu: the fine-grained material that we could collect was concentrated in the small bowl-shaped craters that pocked the surface. Once we realized this, our team set about solving the challenge that Bennu had thrown at us, as we describe in the chapters that follow.

In addition to showcasing the incredible stereo images produced by Brian and Claudia, this book features the contributions of imaging engineer Carina Bennett, who led the painstaking effort to stitch together thousands of images into the global basemap of Bennu and created beautiful mosaics of local landmarks; science editor Cat Wolner, who has overseen all of OSIRIS-REx's academic publications and brought her skills here to ensure the quality for which OSIRIS-REx is known; and planetary geologist Ken Coles, lead author of *The Atlas of Mars* (Cambridge University Press, 2019), who brought his experience in cartography, as well as a fresh eye to help us see Bennu from the point of view of a newcomer.

The OSIRIS-REx mission has been an arduous and rewarding journey. This volume celebrates the whole team's accomplishments in exploring Bennu, uncovering many of its secrets, and advancing our understanding of the early Solar System. We share some of those experiences and insights with the reader to shed light on the fascinating world of asteroid science and exploration.

Dante Lauretta
Tucson, AZ
February 15, 2023

Introduction by Brian May

This book on the *Anatomy of an Asteroid* is absolutely the first of its kind. Never before has an asteroid been so meticulously photographed, studied, and characterized at close quarters. Here, we document the OSIRIS-REx mission to asteroid Bennu, from its inception to its culmination in a successful collection of surface material. As we go to press, the pristine sample is sealed, ready for delivery, as the OSIRIS-REx space vehicle heads back towards Earth.

For me as a partner author, this project has been an amazing journey of discovery. Some years ago I embarked on my first experience of writing with astronomy experts, producing *BANG! — The Complete History of the Universe* — with Sir Patrick Moore and Professor Chris Lintott. Then, as now, I defined my role as a kind of mediator. As we worked our way through every sentence of every page, I played the part of the layman, trying to understand in depth many of the concepts that were new to me, with the motto that if the words and pictures made it clear to me, an inquisitive nonspecialist, then there is a good chance that our work will be understandable to anyone else with a similar keen interest.

The expert role here has of course been played by team leader Professor Dante Lauretta, plus contributions from his whole brilliant team of specialist engineers, navigators, image processors, mineralogists, astrophysicists, and spectroscopists, who together pulled off this amazing feat — a flawless exploratory visit to a celestial object about half a kilometer in diameter, in an orbit which brings it very close to the Earth's path around the Sun.

Because this book is designed to be accessible to all, it is not a chronicle of every moment, movement, or nut and bolt of the mission. For the complete history in every detail we refer you to the references we have included, accessing the many papers published in scientific journals based on the mission findings.

The book includes, as context, a detailed history of the human race's gradual discovery and increasing awareness of asteroids. And, as we follow this story of the exploration of Bennu, our narrative makes use of insights gained from the vast amount of collected data to make inferences as to the nature of asteroids in general.

To me, much of this has been a revelation. When I first joined in this work, I had no idea how extraordinarily important research into asteroids was, and how central it has become to an understanding of how life evolved on Earth. I was actually shocked to realize that I had never clearly asked myself the question of how we could be looking around us at the world we live in, filled with water, air and minerals, including metals, knowing that our planet evolved from an incandescent fireball, a few billion years ago, and not wondering how this could happen. How many of us have focused on that fireball long enough to realize that at such a temperature, all water would be boiled off into space, along with all the gases which presently make up the air we breathe? And presumably all the metals would be molten and sink to the center of the planet, never to be seen again. So how is it that our planet became this pale blue dot, blessed with water and air, plus abundant metals and minerals strewn conveniently on the surface for us to grow food and mine to build our civilization? The notion that these riches can only have been brought to planet Earth by asteroids is actually unavoidable, but it's a shock to our sensibilities. There has been much talk recently, and rightly so, of the danger of an asteroid strike on planet Earth, and the implications of a collision with Bennu are discussed in our text. But who was aware until recently that asteroids, as well as being prime destroyers, may also be prime bringers of life?

Among the many other revelations in this book is the fact that Bennu, probably quite typically, in near-Earth objects, is not a solid body at all, but an accumulation of debris from previous impacts between larger bodies — essentially a pile of rubble only loosely held together by a very weak gravitational field and perhaps a small amount of "stickiness" between the component rocks. At the time of the instigation of the Bennu mission, this was just beginning to be suspected, and it's only recently that I've seen the term "rubble pile" become common parlance — evolving from a mildly insulting epithet into a generally accepted scientific designation for these intriguing and highly delicate bodies. Bennu, in common with Ryugu — a similar asteroid recently visited by the JAXA mission Hayabusa2 — is a spinning-top shaped rubble pile.

We examine in the book the many devices that the OSIRIS-REx craft carried on board, enabling it to make its multidimensional survey of all these characteristics in Bennu. The final dimension is provided by Stereoscopy — side-by-side Victorian-style Brewster format 3-D views. My colleague Claudia Manzoni and I were co-opted to the team soon after launch, to produce three-dimensional images of the explored terrain from the visual data as they were being acquired, with two aims in mind: firstly to document the object in a way that would give the clearest and most beautiful insights into the morphology of the asteroid; and secondly to provide views that would assist the vital process of finding a safe touch-down-and-go site to collect the sample.

We hope that the addition of our stereoscopic views will help this book communicate the intricate anatomy of this now uniquely well-charted Solar System body.

This is the first time such a set of stereo images has been published in book form, so here follows an explanation of how they were created, and, next to the pocket for the OWL Stereoscopic viewer included at the back of the book, you will find instructions on how to optimize your viewing in 3-D.

We at the London Stereoscopic Company are very proud to be the publishers of *Bennu 3-D — Anatomy of an Asteroid*, and hope this book will find its place among the classic reference works of space exploration. And I'm indebted to Dante Lauretta for one of the most challenging collaborations I've ever been privileged to engage in.

Stereoscopic Imaging

A Lightning Explanation

Why did we go to the trouble of making stereoscopic images of Bennu? And what exactly are they designed to achieve? The fact that humans, in common with much of the rest of the animal kingdom, have evolved with not just one eye, but two, separated by about 65 millimeters (two-and-a-half inches), is very significant in evolutionary terms. If we are lucky enough to have healthy vision, at every moment of our waking lives our two eyes communicate two slightly different views of the universe to our brain. Our brains over millions of years have evolved the capability to combine these two slightly different pictures into a single experience — a three-dimensional picture inside our head. We experience a perception of depth, solidity, roundness, shininess, and proximity which is updated every instant without us even thinking about it. This perception was discovered by Englishman Charles Wheatstone in 1832, and is known as "stereopsis". It's an incredibly powerful piece of mental processing which we mostly take for granted in our everyday lives. But Wheatstone's discovery went further than this insight. He realized that if those two different views of our surroundings were captured separately, and subsequently delivered separately to our eyes, the sensation of stereopsis would be recreated, literally in front of our eyes. He predicted that it would be possible to make in-depth captures of scenes rather than the flat pictures which first appeared on the walls of caves at least 60 thousand years ago. The genius of Wheatstone is even more striking when you consider that he made this discovery a few years *before* the invention of photography, by Louis Daguerre and Henry Fox-Talbot.

Sir Charles Wheatstone (1802–75), Professor of Experimental Philosophy at King's College London, the father of stereoscopy, now often referred to as 3-D.

So in this way stereo photography was born, and stereoscopic images became a massive craze in Victorian times, along with the hardware needed to view them, which Wheatstone had christened the "Stereoscope". The first glory days of "Stereoscopy" lasted only about ten years, when it began to be replaced by other crazes, and it only became popular again in the 1950s when it was re-named "3-D" — short for "Three Dimensions". In the 21st century we are all familiar with 3-D films in the cinema. Stereoscopic imaging technology is also widely employed in surgical procedures, and enjoyed in a slightly modified form in virtual reality video games. Over the last 20 years or so, stereoscopy has been increasingly used to interpret data collected by unmanned Solar System missions, mounted by international space agencies such as NASA, ESA and JAXA. We have been proud to be a part of this innovation.

Stereoscopy in space

Why use stereoscopy? In a mission to explore an astronomical body — a planet, moon, asteroid, comet, etc. — the cameras on board the space vehicle are effectively our eyes, because we cannot be there in person. And, just as in everyday life, it's clearly much more informative to use two eyes rather than just one. But it goes a little deeper than that. If we really were gazing out of the window of a spaceship parked a few miles away from Bennu, it would be an awesome sight, but not very stereoscopic! Unless there were some object in the foreground, our two eyes would be receiving almost exactly the same information, because their separation is insignificant compared with the distance of the asteroid from us. So our view would be essentially "flat".

If we are to use stereoscopy to its maximum effect, to instinctively enjoy such objects in depth, we need to modify the image capture process. In order to get significant parallax differences between the images for our left and right eyes, we'll need to place our cameras much further apart — on the same scale as our distance from the object — in this case, a few miles. This increased separation between camera positions gives us what has been traditionally called a long "baseline", and the resulting exaggerated stereo image is known as a "hyperstereo". Almost all the stereo images in this book are hyperstereos. It's as if we became giants, or as if Bennu were shrunk to a rock which fits in our hand, and could be viewed up close. But how do we position those two viewpoints for the two cameras several miles apart? Assuming we have only one spacecraft, we actually can't — at least not simultaneously. But if we wait a while between exposures, the vehicle will have moved relative to the asteroid, and we just need to wait until a suitable distance has been covered for a usable "baseline". In practice, we are generally compiling stereos after the event, so, once the whole collection of images has been downloaded from the craft, it's a matter of carefully sifting through the data to find two such suitable images to pair up.

To make the kind of stereo images you will find in this book, we look for two versions of any given scene captured from viewpoints just far enough apart to give us the parallax we need for a viewable hyperstereo. Of course we are also looking for the clearest images, with the best resolution. Often both the left and right images have to be preassembled by seamlessly stitching together two or more of the raw "plates" delivered by the OSIRIS-REx craft's cameras. Ideally we would like one more thing too — the same illumination in both images, so the shadows match up (we then call our stereo "synchronous"). But only very seldom is this possible.

Alignment

The two component images of the stereo pair are adjusted to match up in scale, brightness, and color. The images are then rotated and aligned so that the all-important parallax shifts between the left and right views are horizontal in the view. This last requirement is vital, because the optical processing inside our brains is not equipped to deal with a vertical component of parallax — such discrepancies just cause confusion and headaches. Part of this process is ensuring that objects in the left and right frames are exactly aligned vertically. Horizontally, the components are adjusted so that, viewed in the stereoscope, objects in the view sit comfortably behind the "stereo window" formed by the edges of the images. If they come too far forward in the depth dimension (sometimes referred to as "Z-plane"), they will cause uncomfortable violations of the stereo window, and if they are too far back, they will draw the convergence of the axes of our eyes too far apart, and the stereo will not be viewable without discomfort. Our eyes are quite good at converging on an object close-up, but as soon as a *divergence* of their axes is called for, our image processing systems rapidly fail!

Shadow adjustments

After all this is done, there is usually one more important step to take. In the time interval between the two exposures, the space vehicle has moved, giving our required long baseline between shots, but in this time the asteroid has also rotated somewhat, introducing another component to the resulting baseline. This is fine, but unfortunately the secondary effect of this rotation is that Bennu's orientation relative to the Sun will have changed too, so all the shadows will fall in different ways, creating serious flaws in the stereoscopic image. Sometimes the anomalies will make the shadows appear to jump out of the plane of the picture, or else give the appearance of a chasm which is not there, and in some cases the stereo becomes impossible to view. So unless we've been lucky enough to secure a synchronous pairing, in which such problems are vanishingly small, the shadow problem has to be dealt with. It's possible that in the future, an AI solution will be found. But as of now, the only method we know of to fix the shadows is to manually map them from one side to the other, taking account of the elevation and shape of the terrain that the shadows fall on. Each shadow from every rock has to be adjusted in this way, until all the disparity is eliminated. This is necessarily a long and labor-intensive process, but the results justify the effort!

The feature Gargoyle Saxum on the surface of Bennu, stereo aligned and adjusted for levels, but not shadow-corrected.

The same stereo, in final form, after the shadow work.

At the end of all this, the stereo image can be further cropped to zoom in on a detail under scrutiny. In a book, the stereo must be printed on the page at a scale carefully chosen to make sure the distance between the left and right images of any given object (especially at infinity) is never much more than the separation of our eyes. There is a little latitude because the OWL stereo viewer is designed to help convergence, its lenses being mounted farther apart than our eyes, to introduce a prismatic effect. So generally the separation between infinities in any stereo pair of images is kept to a maximum of about 75 millimeters (three inches). We trust this will ensure that nobody gets headaches from this book!

Viewing in 3-D

To enjoy the maximum effect of the stereo pictures, please refer to the instructions beside the viewer included at the back of the book.

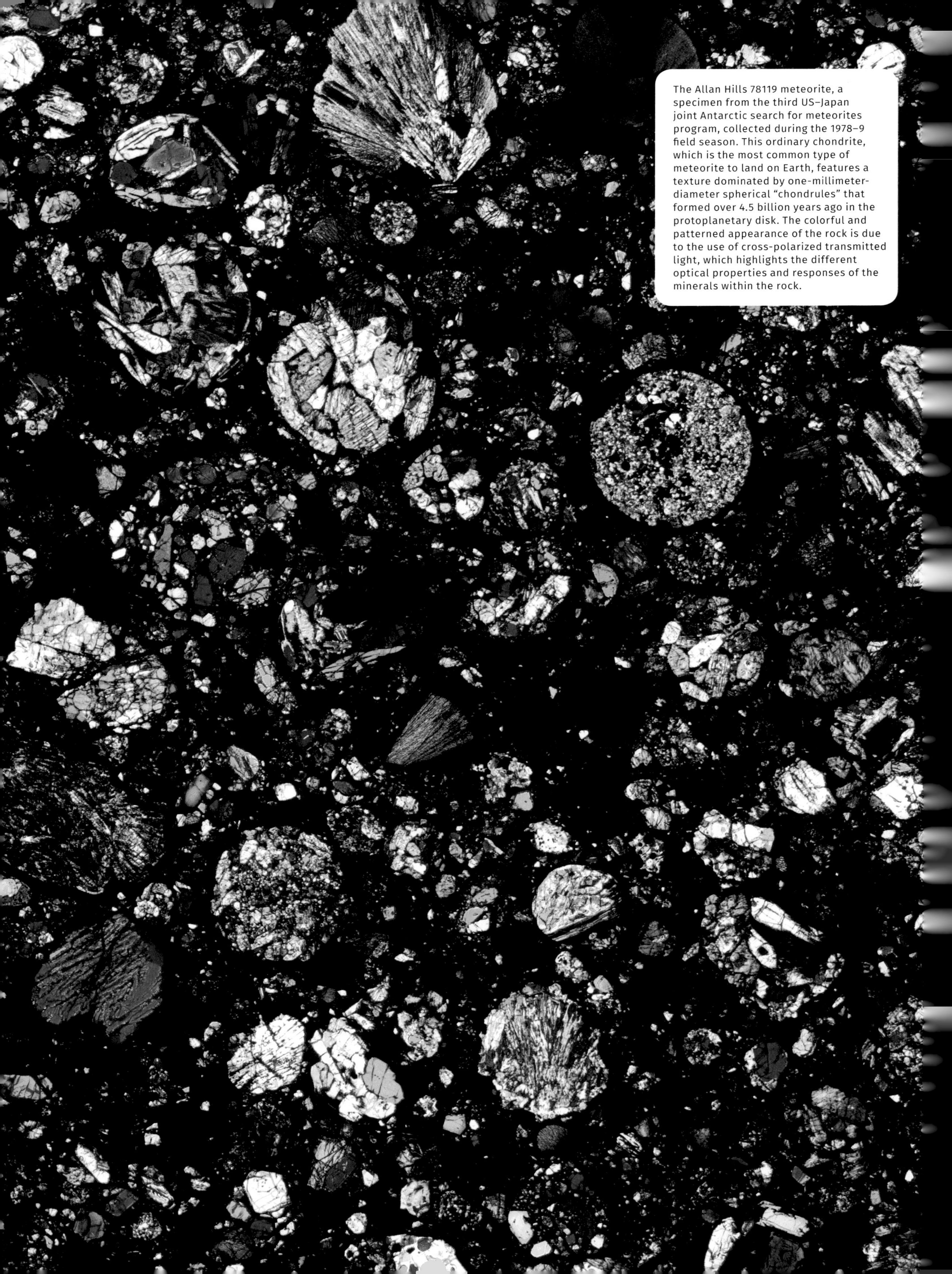

The Allan Hills 78119 meteorite, a specimen from the third US–Japan joint Antarctic search for meteorites program, collected during the 1978–9 field season. This ordinary chondrite, which is the most common type of meteorite to land on Earth, features a texture dominated by one-millimeter-diameter spherical "chondrules" that formed over 4.5 billion years ago in the protoplanetary disk. The colorful and patterned appearance of the rock is due to the use of cross-polarized transmitted light, which highlights the different optical properties and responses of the minerals within the rock.

Chapter 1

All About Asteroids

A little over 4.5 billion years ago, in the Milky Way galaxy, a giant molecular cloud collapsed into a spinning disk of gas, ice, and dust. Most of the mass was concentrated at the center of this disk, forming a young protostar, which would grow to become the Sun. As the cloud contracted, high temperatures vaporized the ice and dust. The disk cooled over the next several million years, and mineral grains condensed like snowflakes, producing fine and fluffy particles suspended in the nebular gas. This material was swept up and accreted to form tiny planets, known as planetesimals, some of which, in turn, collided and coalesced with each other to form planets.

Asteroids are planetesimals left over from this time that never became part of a planet. Even today, they continue to collide, generating millions of fragments of rock and dust. In 1661, the English natural philosopher Joshua Childrey noticed this phenomenon as a diffuse cone of light in the western sky at the end of twilight, and in the eastern sky just before dawn. Italian-born French astronomer Giovanni Domenico Cassini in 1685 suggested, correctly, that dust particles in orbit around the Sun were the source of this "zodiacal light."

Some asteroid fragments end up on Earth-crossing orbits. When they collide with the Earth, surviving their passage through the atmosphere and landing on the surface, they are called meteorites.

Meteorites have long fascinated humanity. Historically many meteorites were venerated as divine messengers, including the Stone of Delphi in ancient Greece, the holy rock of Hadschar al Aswad in the Kaaba in Mecca, and the stone from which the ancient "Iron Man" Buddhist sculpture was carved. Meteorites have been revered by indigenous peoples in North America, Greenland, and Australia. Containing rare and precious metallic iron and nickel, these rocks were literally gifts from heaven.

As samples of asteroids, meteorites are also gifts to scientific knowledge. But before the late 18th century, their value in this respect went unrecognized in Western Europe. The scientific consensus was that fragments of rock and metal did not fall from the sky, and no objects smaller than a planet existed in space beyond the Moon. Even though there were multiple eyewitness reports of meteorite falls, starting with the Ensisheim fall in 1492, the scientific community dismissed them as folk tales. In 1794, Ernst Chladni proposed that fragments of stone and iron frequently enter the atmosphere from space, forming meteors along the way. As Ursula Marvin reports in her 1996 review, Chladni was severely criticized by his colleagues for taking "gross liberties with the laws of physics."

Fortunately for Chladni, four major meteorite falls occurred in populated areas in Europe between 1794 and 1798. In 1802, Edward Howard and colleagues published their discovery that the metal in meteorites includes large amounts of nickel, an element that is extremely rare at the surface of Earth. Their results at last persuaded most leading European scientists that fragments of rock and metal routinely fall from the sky. Any remaining skeptics were defeated in April 1803, when thousands of meteorites were witnessed to fall at L'Aigle in Normandy, all of which contained abundant nickel.

Over the next 150 years, the scientific community would go on to evaluate — and abandon — multiple ideas for the origin of meteorites, including formation within Earth's atmosphere (hence the shared etymological root with "meteorology"), ejection by lunar volcanoes, and delivery from interstellar space.

Parallel to the rise of meteorite science, asteroids began to be noticed in the night sky. In 1801, Giuseppe Piazzi observed an object in space that he thought was a new planet, but it was much smaller than all other planets — smaller in volume even than the Moon, by a factor of 55. Piazzi named the object Ceres, after the Sicilian goddess of grain. Because of its small size, the scientific community dubbed it a minor planet, a term that is still in use. It orbits the Sun between Mars and Jupiter at an average distance of 2.77 astronomical units (AU). One AU is the average distance between Earth and the Sun.

Between 1801 and 1808, astronomers identified three more so-called minor planets between Mars and Jupiter, which they named Pallas, Juno, and Vesta. As reported in Brewster's 1811 astronomy textbook, these discoveries made astronomers lament that they "destroy the harmony of the Solar System," a reference to the apparent pattern in planetary orbits known as the Titius-Bode Law (or just Bode's Law). The discovery of Uranus in 1781, with an orbit consistent with this law, had motivated the search for an object with Ceres' orbit. A fifth object in this same region was discovered in 1845. Named Astraea, it reignited scientific interest in this class of celestial objects, which by that point had come to be called asteroids. Since then, asteroids have been discovered almost every year. It soon became obvious that a massive cloud of asteroids exists between Mars and Jupiter, which is now known as the main asteroid belt.

The debate over the origin of meteorites was settled in the mid-20th century. Scientists analyzed meteorite flight paths through the atmosphere to trace their trajectories back to their "parent bodies," another term that remains in scientific use. This work provided conclusive evidence that meteorites originated in the main asteroid belt.

The pace of asteroid discoveries has since accelerated because of dedicated amateur astronomers and, beginning in 1985, systematic telescopic surveys. To date, over one million asteroids have been discovered. As of February 2023, the discoveries of 620,000 numbered minor planets (those with well-defined orbits) are credited by the Minor Planet Center to more than 1,000 astronomers and 250 observatories, telescopes, or surveys across the globe, with more discovered every day.

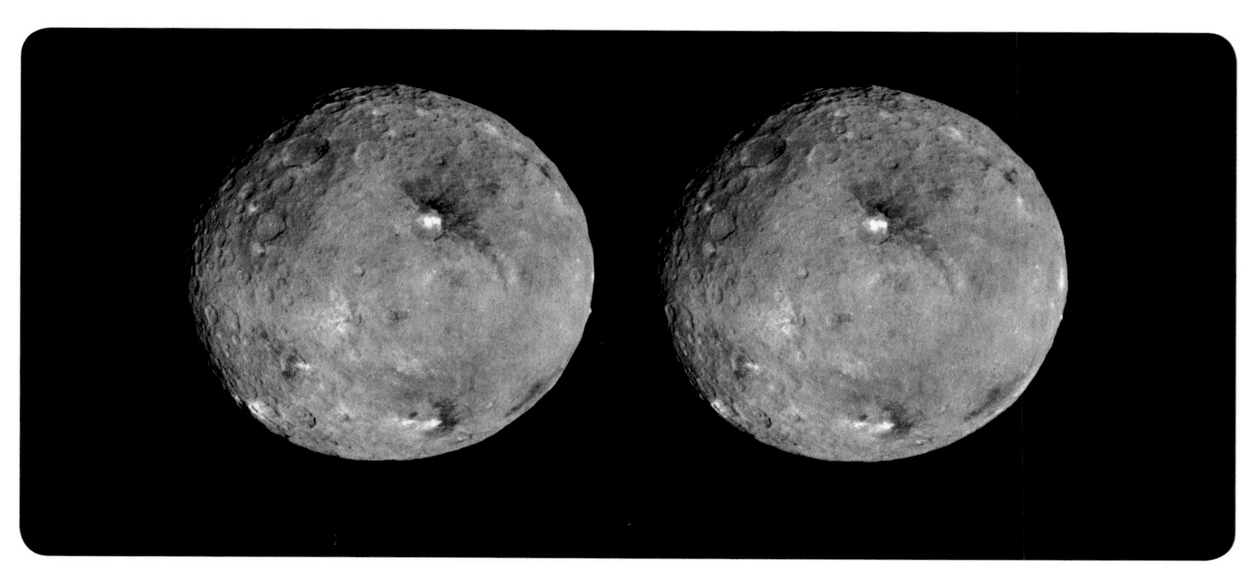

Ceres, the largest asteroid in the main belt, about 940 km (590 miles) in diameter.
Stereo image from data returned by the Dawn spacecraft.

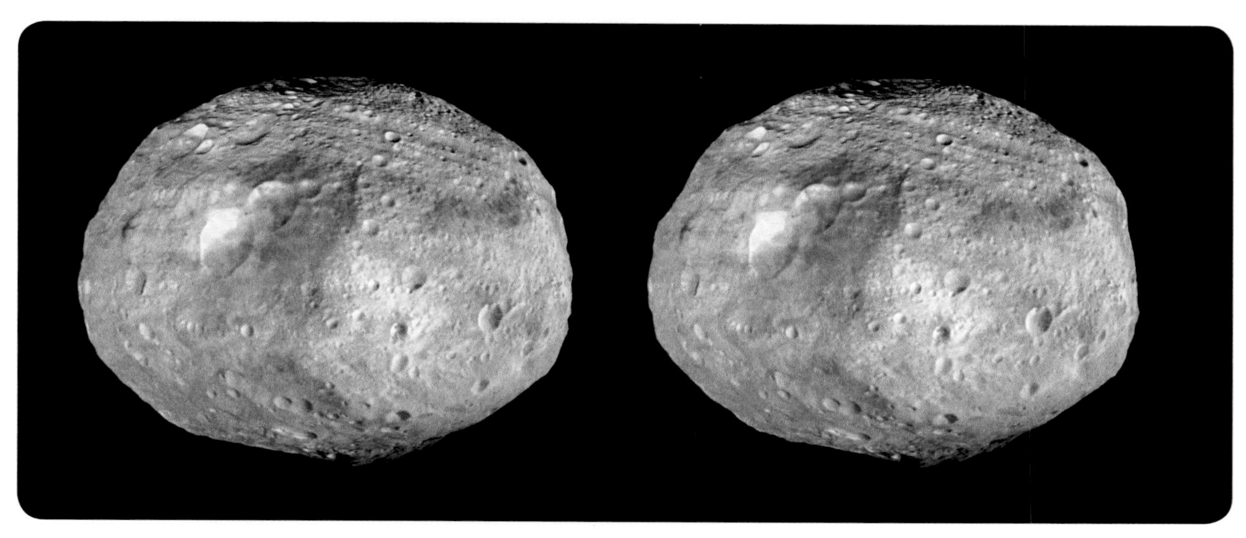

Vesta, another large asteroid in the main belt, about 525 km (326 mi) in diameter, also from Dawn.

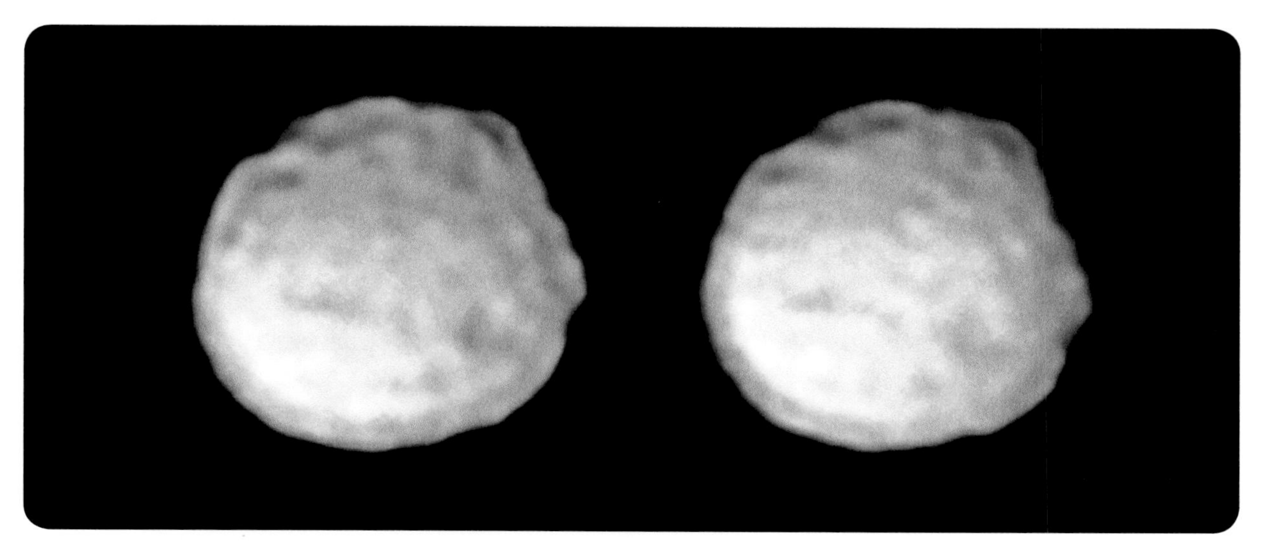

Perhaps the world's first stereo view of asteroid Pallas, using images from the VLT (Very Large Telescope) at Paranal, Chile. There has never been a spacecraft mission to resolve the surface of Pallas, but this image illustrates the power of ground-based telescopes. The resolution is good enough for us to be sure that Pallas is not a rubble pile.

Timeline of asteroid exploration

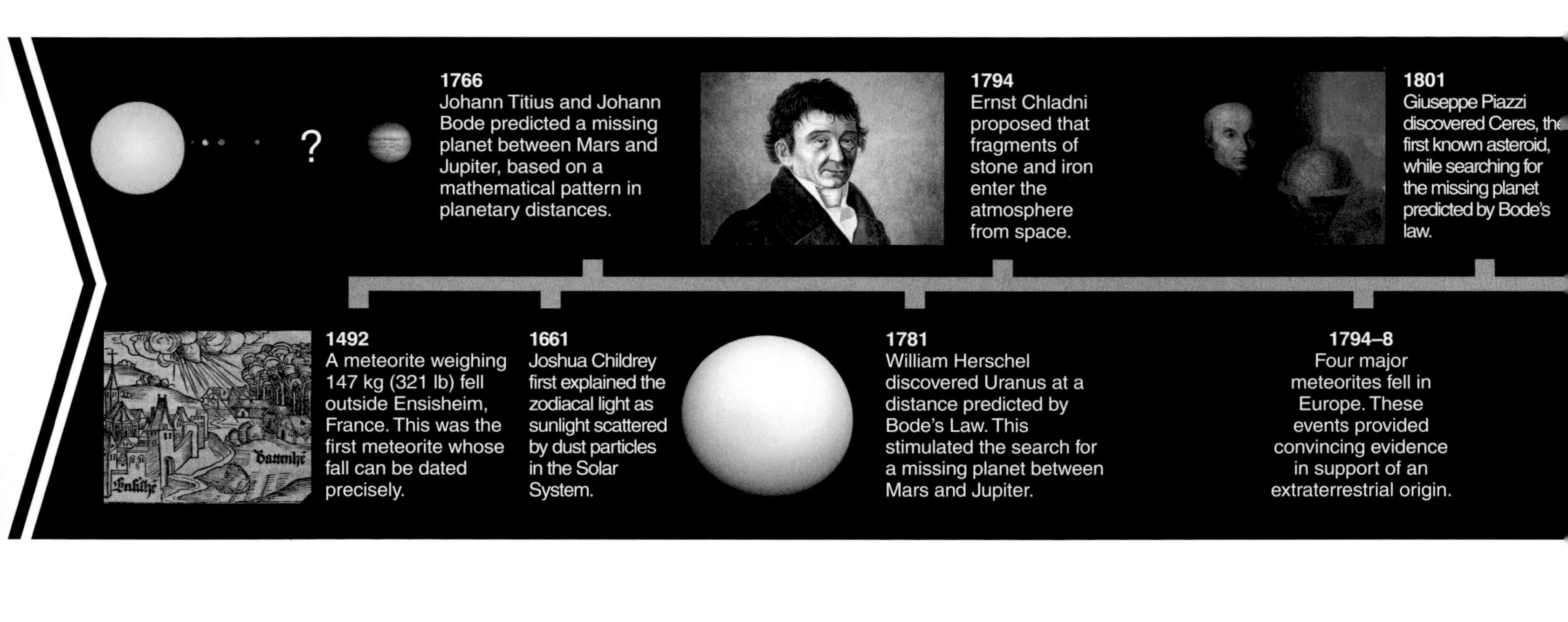

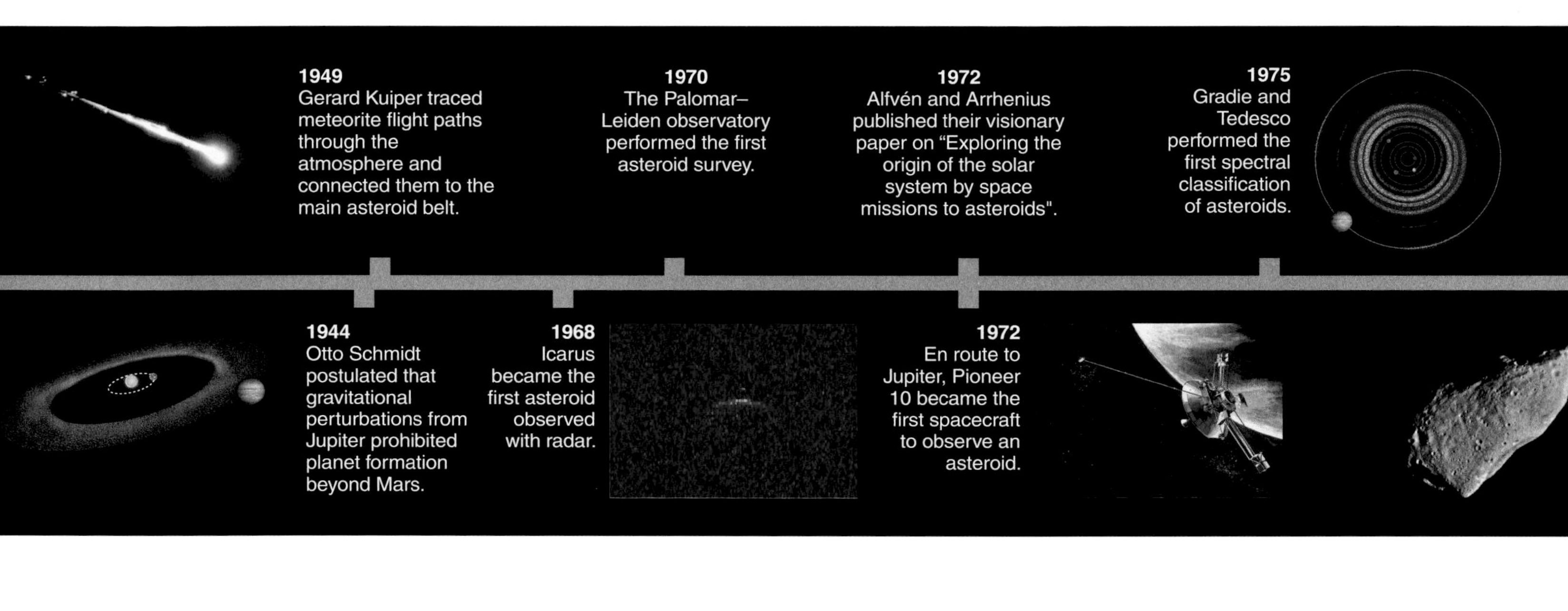

1802
Edward Howard discovered that meteorites include large amounts of nickel, an element that is rare on the surface of the Earth.

1845
After a 37-year slump, a fifth asteroid was discovered – Astraea. This discovery reignited interest in asteroid science.

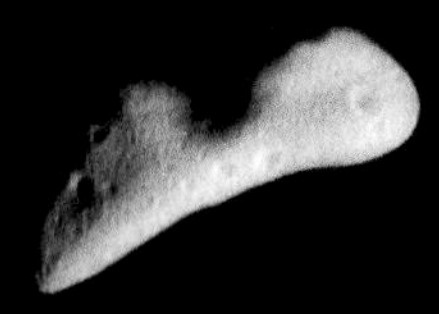

1898
Witt and Linke discovered Eros, the first known near-Earth asteroid.

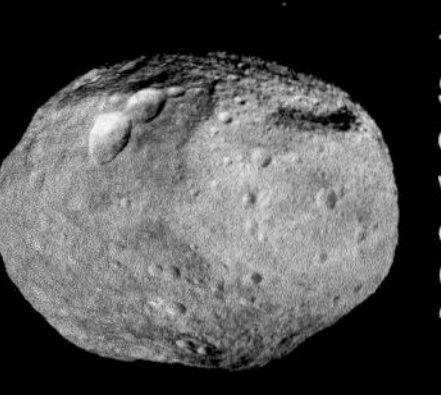

1801–8
Soon after the discovery of Ceres, three objects were found in similar orbits, making it clear that Ceres was one of many "minor planets."

1803
A shower of thousands of stony meteorites fell at l'Aigle in Normandy, removing any doubt that rocks fall from space.

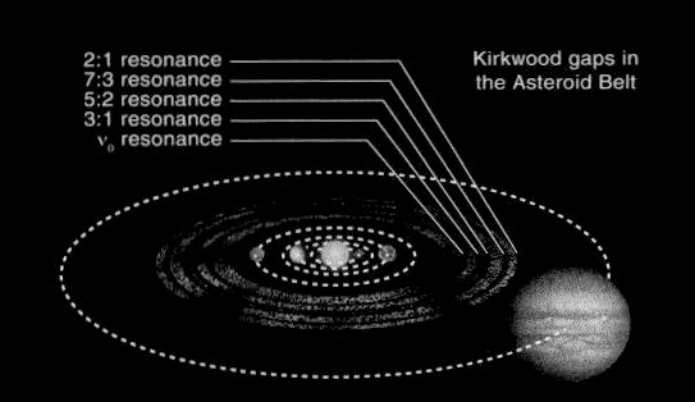

1867
Daniel Kirkwood identified gaps in the distribution of asteroid orbits. These gaps result from the dominant gravitational influence of Jupiter.

1993
Galileo spacecraft flew by Ida and discovered the first asteroid satellite.

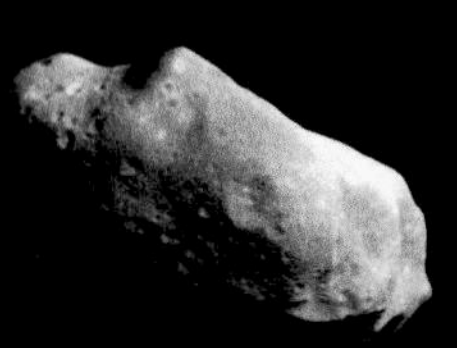

1997
The NEAR Shoemaker spacecraft flew by Mathilde, the first carbonaceous asteroid observed by a spacecraft.

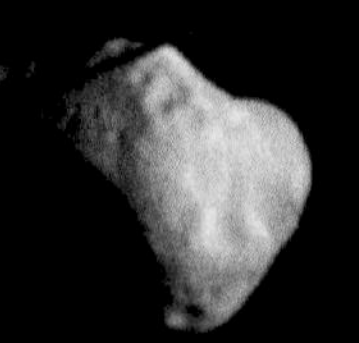

1999
Bennu and Ryugu were discovered by the LINEAR hazardous asteroid survey.

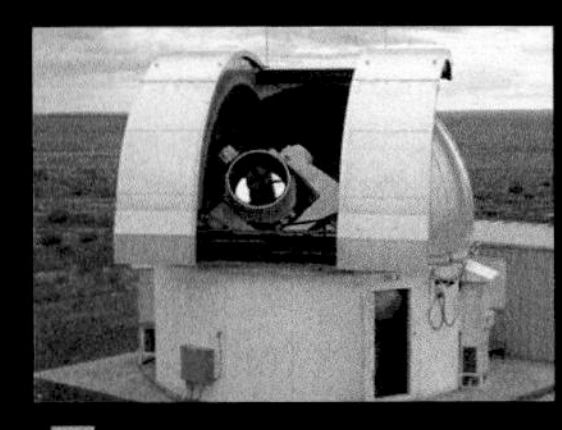

2000
The Cassini spacecraft flew by Masursky en route Saturn.

1991
The Galileo spacecraft obtained the first resolved image of an asteroid – Gaspra – on its way to Jupiter.

1995
The Spacewatch survey in Arizona came online to identify potential Earth impactors.

1998
NEAR Shoemaker arrived at Eros and became the first spacecraft to orbit and land on an asteroid.

1999
The Deep Space 1 spacecraft flew by Braille.

2002
The Stardust spacecraft flew by Annefrank.

2011
The Dawn spacecraft arrived at Vesta.

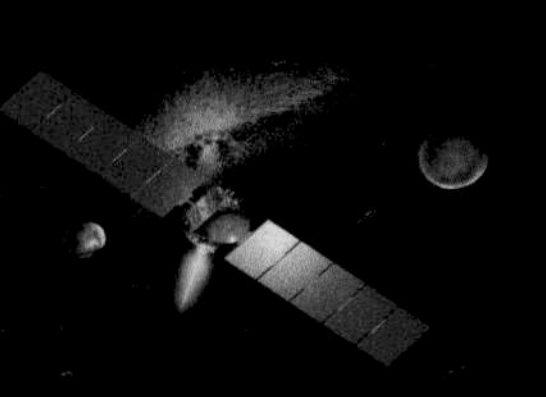

2019
The Hayabusa2 spacecraft collected samples of Ryugu.

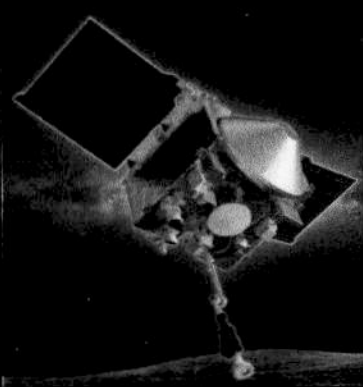

2020
OSIRIS-REx collected samples of Bennu.

2023
OSIRIS-REx returned samples of Bennu to Earth.

2015
Dawn arrived at Ceres, making it the first spacecraft to orbit two asteroids.

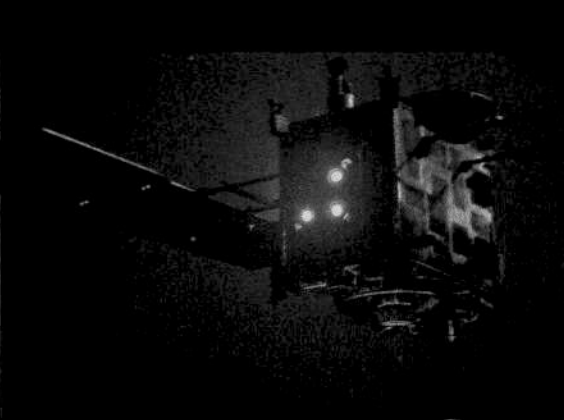

2020
Hayabusa2 returned samples of Ryugu, confirming the relationship between C-type asteroids and carbonaceous meteorites.

The asteroid alphabet

The early asteroid surveys of the 1980s were quickly followed with techniques to understand their diversity. To infer the composition of such distant objects, scientists looked at their optical and spectral properties — that is, how they interact with light. The classification system for asteroids has evolved over time, but it is generally based on the total light reflected from the asteroid's surface (albedo), changes in reflectivity across different wavelength bands (spectral "color" and "slope"), and, for the brightest objects only, spectra across a wide wavelength range that are detailed enough to act as an identifying fingerprint.

The resulting taxonomic types are colloquially dubbed the asteroid alphabet:

- **B- and C-types**: Asteroids with low albedos and nearly flat, featureless spectra. They are most abundant in the outer main belt. The B stands for "blue," describing a negative spectral slope. The C stands for "carbonaceous" (relatively carbon-rich), a property thought to be common to both types.
- **D-types**: Asteroids with low albedos, relatively featureless spectra, and a very steep positive ("red") spectral slope. They are abundant near Jupiter; hence D for "distant" asteroids
- **S-types**: Asteroids with moderate albedos and a spectral absorption feature in the near-infrared. They are brighter than the C-types because they lack carbon. They are most abundant in the inner main belt. The S stands for "stony."
- **V-types**: Asteroids with higher albedo in the inner main belt whose spectra show an affinity with volcanic rocks. The V stands for asteroid Vesta, which is the type specimen and source of the V-type asteroids (also known as vestoids).
- **X-types**: Asteroids with flat, featureless spectra, making it very difficult to infer their composition. The X stands for the unknown. They have been subclassified into three types: E-types have high albedos, M-types have mid-range albedos, and P-types have low albedos. These asteroids occur throughout the main belt.

Bennu, destination of the OSIRIS-REx mission, is a B-type asteroid.

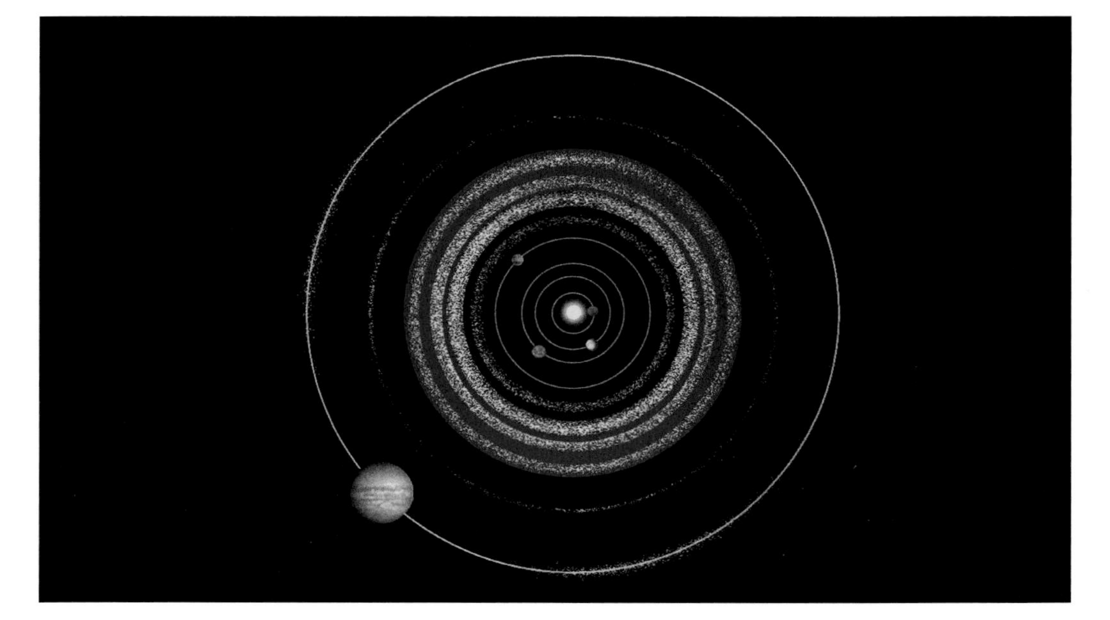

The stratification of the asteroid belt. The gaps occur where asteroid orbits are in resonance with Jupiter. The color coding indicates the compositional variation with green indicating S-types; blue, B- and C-types; and red, D-types. V-types are not abundant enough to show.

Asteroids in motion

In addition to classification by their spectral properties, asteroids are also grouped into "families" on the basis of similarity in their orbits. Members of a given asteroid family are typically fragments of varying sizes from the same giant asteroid collision. Some are clusters of fragments that have gravitationally coalesced; these typically small asteroids, of which Bennu is an example, are known as "rubble piles."

The gravitational pull of the giant planets Jupiter and Saturn are the main forces altering the orbits of asteroids. However, for small asteroids, less than 10 kilometers (6.2 miles) in diameter, other factors such as sunlight can play a significant role. Asteroids receive most of their sunlight at noon, local solar time. Because of thermal inertia (resistance to heating and cooling), there is a lag between when the maximum energy from the Sun is received at noon and when this energy is released back into space as heat later in the afternoon. The emitted thermal radiation carries momentum, and thus it imparts a miniscule force that has a long-term effect on the asteroid's orbit. This force is known as the Yarkovsky effect.

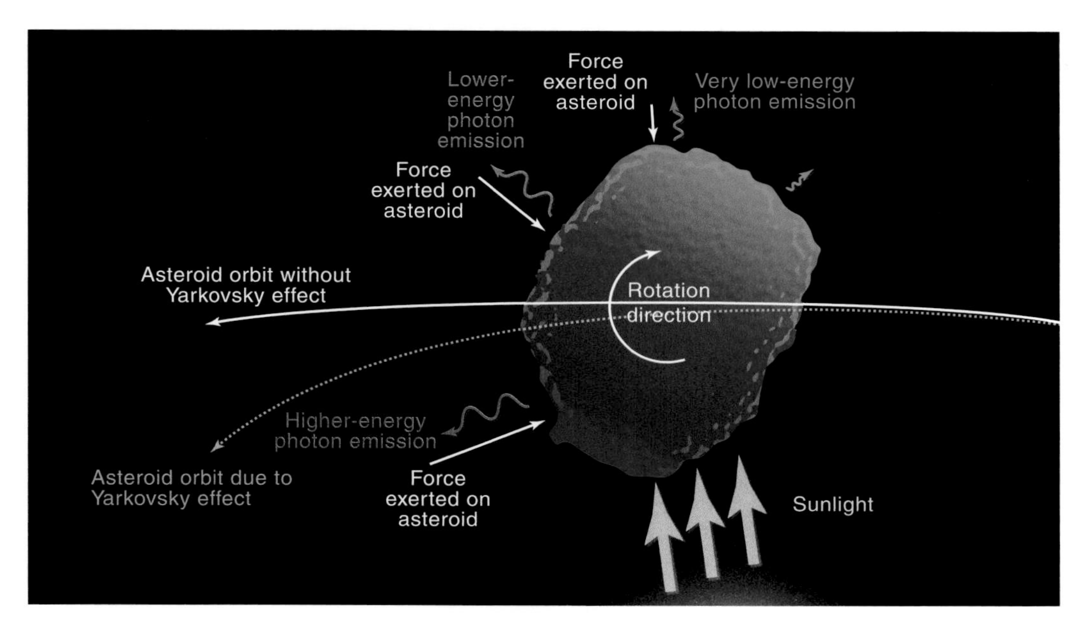

The Yarkovsky effect.

Another phenomenon, called the YORP effect (named after the four scientists who contributed to the theory: Yarkovsky, O'Keefe, Radzievskii, Paddack), also results from the absorption of sunlight and reemission of thermal radiation. The YORP effect can create a torque that modifies the spin of small asteroids — driving the asteroid's rotation axis to be perpendicular to the sunlight and causing the rotation rate to increase or decrease. Hence, the YORP effect can change the way that the Yarkovsky effect operates.

These thermal effects considerably influence the fate of small asteroids. The solar-powered thrust of the Yarkovsky effect moves small asteroids across the main belt. Those with prograde rotation (spinning in the same direction as their orbit) emit thermal radiation behind them, increasing their orbital velocity and hence their average distance from the Sun (semi-major axis). Conversely, those with retrograde rotation (spinning in the opposite direction to their orbit), including Bennu, emit their heat in front of them, which slows them down and pushes them inward — onto trajectories leading toward Earth. Understanding these effects is critical to predict the probability of an asteroid impacting our planet.

Our asteroid neighbors

In the main belt, asteroids are not evenly distributed between Mars and Jupiter; there are gaps where the belt has been sculpted by orbital "resonances" with Jupiter and Saturn. These resonances occur where asteroid orbital periods are in an exact integer ratio with one of the giant planets.

When small rubble-pile asteroids form after giant collisions, the Yarkovsky effect causes them to migrate towards these resonances, where their orbits become unstable and the giant planet's gravity flings them into the inner Solar System.

Thus, there is a steady supply of asteroids tumbling into Earth's vicinity from the main belt. If their orbits cross Earth's, they are known as a near-Earth asteroid (NEA). More than 30,000 NEAs have been discovered.

NEAs are categorized into four types based on their orbital elements, each named for the type-defining asteroid:

- **Atira-types:** NEAs whose orbits are entirely within Earth's.
- **Aten-types:** Earth-crossing NEAs with semi-major axes smaller than Earth's.
- **Apollo-types:** Earth-crossing NEAs with semi-major axes larger than Earth's; Bennu belongs to this category.
- **Amor-types:** Earth-approaching NEAs with orbits exterior to Earth's but interior to Mars'.

Missions to asteroids

In the early years of space exploration, asteroids were glimpsed distantly by spacecraft en route to other destinations, such as Jupiter and Saturn. With the Near-Earth Asteroid Rendezvous (NEAR) Shoemaker mission to Eros in the late 1990s, asteroids became primary targets of exploration, and the first landing on an asteroid surface was made.

In the early 21st century, the Hayabusa mission brought back the first asteroid samples — grains from an S-type NEA called Itokawa — and the Dawn mission mapped the surfaces of two large main-belt asteroids, V-type Vesta and C-type Ceres. We are in a time of exponential growth in our understanding of asteroids via the close encounters, high-resolution observations, and samples returned by modern missions such as Hayabusa2 to the C-type NEA Ryugu, and OSIRIS-REx to the B-type NEA Bennu.

Asteroids Encountered by Space Missions

Encounter year(s)	Asteroid	Mission	Encounter description
1972	(307) Nike	Pioneer 10	Incidental flyby en route to Jupiter from a range of 8.85 million km (5.5 million mi)
1991	(951) Gaspra	Galileo	Incidental flyby en route to Jupiter from a range of 1,604 km (997 mi)
1993	(243) Ida	Galileo	As above, from 2,410 km (1,500 mi)
1997	(253) Mathilde	NEAR Shoe-maker	Flyby with closest approach of 1,212 km (753 mi)
1998–2001	(433) Eros	NEAR Shoe-maker	Flyby from 3,827 km (2,378 mi) in lieu of aborted orbit insertion in 1998; one year of orbit in 2000–1, ending with a controlled descent, the first "landing" on an asteroid
1999	(9969) Braille	Deep Space 1	Flyby at 26 km (16 mi) that returned two images and three infrared spectra
2000	(2685) Masursky	Cassini	Incidental flyby en route to Saturn from a range of 1.5 million km (0.9 million mi)
2002	(5535) Annefrank	Stardust	Flyby at 3,079 km (1,913 mi) that imaged the asteroid using the navigation camera
2005	(25143) Itokawa	Hayabusa	Rendezvous that performed station keeping and contacted the surface twice, collecting the first sample of asteroid material (less than a milligram)
2008	(2867) Šteins	Rosetta	Flyby with a closest approach distance of 800 km (500 mi) that acquired data from 15 different instruments
2010	(21) Lutetia	Rosetta	Flyby with a closest approach distance of 3,170 km (1,970 mi)
2011–12	(4) Vesta	Dawn	Orbited from a range of altitudes that varied from 210 to 2,750 km (130 to 1,709 mi). Performed global mapping using its framing camera, a visible–near-infrared spectrometer, and a gamma-ray spectrometer
2012	(4179) Toutatis	Chang'e 2	Flyby with a closest approach distance of 770 m (0.5 mi) that collected more than 400 useful images
2015-18	(1) Ceres	Dawn	Orbited from a range of altitudes that varied from 35 to 1,480 km (22 to 920 mi). Follow-on target from the Vesta encounter
2018–19	(162173) Ryugu	Hayabusa2	Rendezvous mission that performed station keeping, deployed a small lander and two rovers (a first for asteroid exploration), and collected 5 gm of surface and subsurface samples
2018–21	(101955) Bennu	OSIRIS-REx	Performed a series of flybys and orbits at ranges from kilometers to hundreds of meters, culminating with a touch-and-go maneuver to collect material from the surface and subsurface (hundreds of grams)

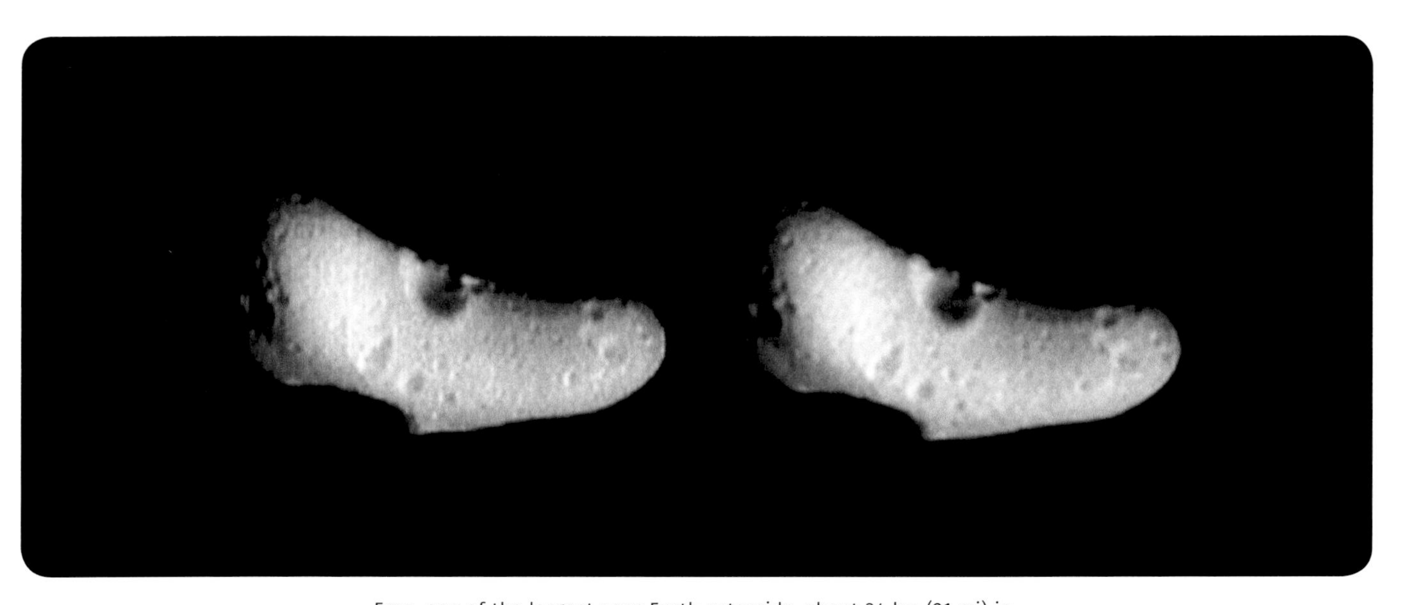

Eros, one of the largest near-Earth asteroids, about 34 km (21 mi) in longest dimension, from images by the NEAR Shoemaker spacecraft.

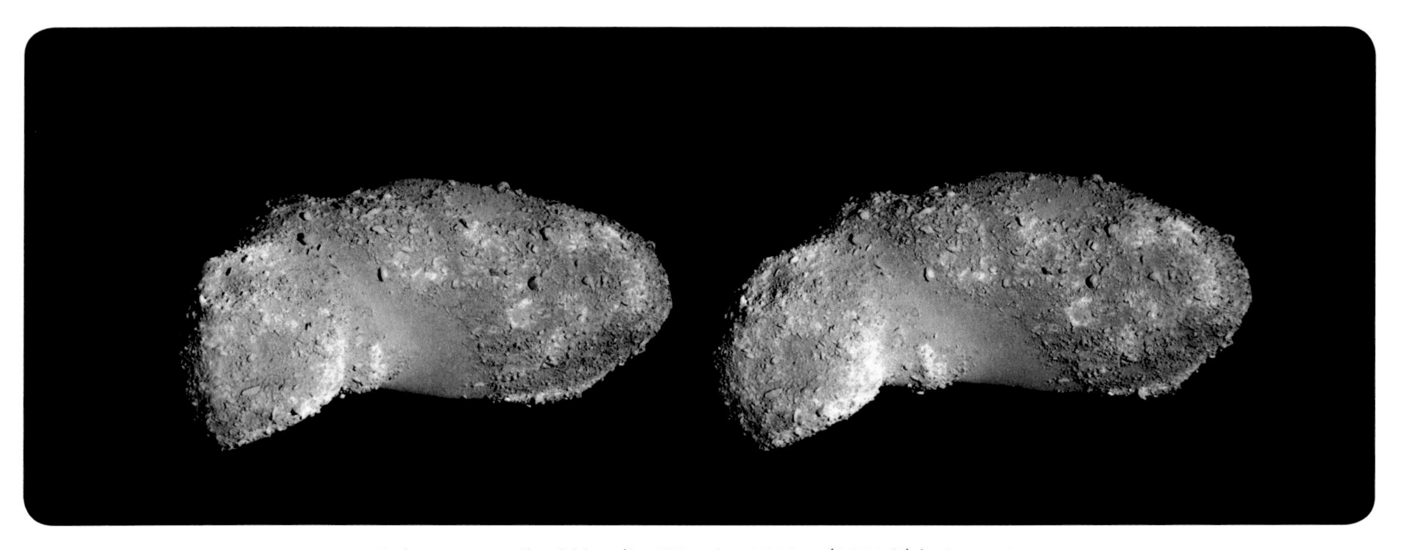

Itokawa, a small rubble-pile NEA, about 330 m (1,082 ft) in longest dimension, from images captured by the Hayabusa spacecraft.

Ryugu, a rubble-pile NEA about 900 m (3,000 ft) in diameter, from images by the Hayabusa2 mission.

1969: A meteorite bounty

The study of meteorites underwent a renaissance in 1969, when three unrelated events occurred: the fall of the Allende meteorite in Mexico, the fall of the Murchison meteorite in Australia, and the recognition of large meteorite deposits in Antarctica.

In the early morning of February 8, 1969, a massive fireball was observed in the skies above the Chihuahuan village of Pueblito de Allende, Mexico. The fireball, now known as the Allende meteorite, was accompanied by multiple loud sonic booms as it exploded into pieces during its passage through the atmosphere. The pieces landed over an area extending more than 300 square kilometers (116 square miles), with collected specimens ranging from one gram to 110 kilograms (243 pounds), and a total recovered mass of more than 2,000 kilograms (4,409 pounds). Later that year, on September 28, 1969, the Murchison meteorite entered Earth's atmosphere in Victoria, Australia. It also fragmented as it passed through the atmosphere, and many samples were recovered, totaling more than 100 kilograms (220 pounds).

A piece of the Allende meteorite, 4 cm (1.6 in) in longest dimension. The bright white inclusions are rich in calcium and aluminum. They are the oldest known mineral grains and define the age of the Solar System.

A piece of the Murchison meteorite, 4.5 cm (1.8 in) in longest dimension. Its dark color reflects the high carbon concentration. The carbon is rich in organic molecules, similar to those that may have seeded the Earth with prebiotic compounds.

ANSMET scientists (including Professor Lauretta, far right) hunting for meteorites in Antarctica's MacAlpine Hills during the 2002–3 field season.

The fragments of Allende and Murchison were studied in laboratories around the world, including the state-of-the-art facilities that NASA had built to study the lunar rocks returned by the Apollo missions. The Allende meteorite was found to contain bright white inclusions rich in calcium and aluminum, with compositions similar to the re-entry tiles used on NASA's space shuttles. These inclusions were the oldest solid objects ever measured: their radioisotope ages are still used to define the 4.567-billion-year age of the Solar System.

The Murchison meteorite was quickly recognized as having high concentrations of extraterrestrial organic compounds. When a team of researchers discovered the first evidence of extraterrestrial amino acids in 1970, the idea took hold that meteorites from asteroids may have delivered the building blocks of life to Earth.

After more than 50 years, Allende and Murchison remain two of the most studied meteorites. They may be representative of the rocks on Bennu: both are carbonaceous and were geologically altered by ice melting in their parent bodies (Chapter 3).

The recognition of large concentrations of meteorites in Antarctica was made by an inland survey team of the Japanese Antarctic Research Expedition (JARE) in 1969. On December 21 of that year, members of the survey team collected three stones from the surface of the ice sheet in the southeastern Yamato Mountains. The team realized that these rocks might be meteorites, prompting them to search the exposed blue glacial ice and collect other candidates. After their return to Japan, the nine stones they collected in Antarctica were all identified as meteorites. This discovery prompted annual systematic searches by both Japanese and American expeditions. JARE expeditions have collected over 17,000 meteorites. The US-led Antarctic Search for Meteorites program (ANSMET) has operated almost every year since 1976, recovering more than 22,000 specimens. This regular supply of new samples has allowed meteoritics to be a vibrant and active area of scientific research into the formation and evolution of our Solar System.

Linking meteorites and asteroids

More than 50,000 meteorites have been found on Earth, most (39,000) of them in Antarctica. The majority (99.8%) are thought to have come from asteroids and are more than 4.5 billion years old. The remaining 0.2% come from the Moon or Mars.

The asteroidal meteorites can be categorized in many ways. Classically, meteorites were divided into stony, stony-iron, and iron categories. Today there are more than 50 different meteorite groups, based on the composition, mineralogy, and petrography of the samples. At the highest level, they can be usefully divided into three broad categories: chondrites, achondrites, and iron meteorites.

Chondrites are typically characterized by the presence of chondrules — round particles that solidified from melt droplets in the protoplanetary disk. The most "primitive" chondrites (those least altered from their original state) are sedimentary rocks that swept up various amounts of metal, silicates, ice, and organic compounds as they moved through the disk, forming the first asteroids. Different chondrite groups record various levels of geologic alteration in their parent asteroids' interiors. They show that asteroids experienced internal heating, ranging from 25 to 1,250 °C, owing to the decay of short-lived radioactive elements. In carbonaceous asteroids with ice, this led to melting, followed by hydrothermal alteration by liquid water. Those that accreted "dry" (without ice) experienced thermal metamorphism as recorded in the ordinary and enstatite chondrites. These processes altered the mineralogy and the texture of the material but left the bulk chemical composition intact.

Chondrites are divided into three classes: carbonaceous, enstatite, and ordinary. Ordinary chondrites, as their name implies, are the most abundant type of meteorite on Earth, constituting more than 85% of observed falls. They are composed of metal and silicate minerals. They accreted without ice and therefore experienced dry thermal

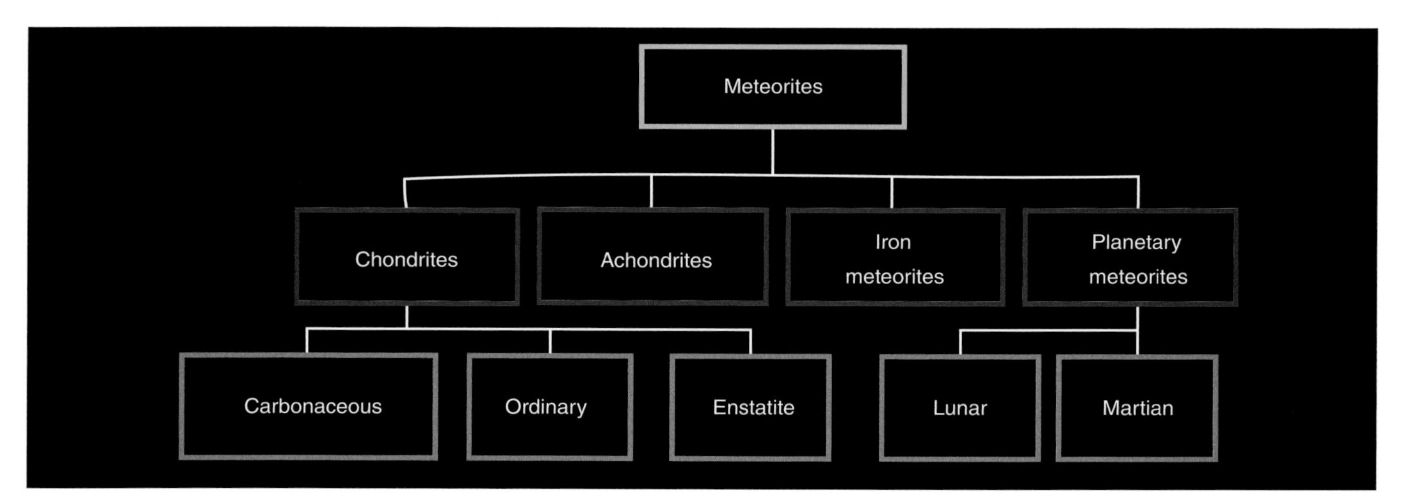

The classification of meteorites.

metamorphism. The enstatite chondrites are named after their most abundant mineral, the magnesium-rich pyroxene enstatite. The enstatite chondrites formed under chemically reducing conditions, resulting in distinct mineralogies. The carbonaceous chondrites have experienced a range of geologic histories, with a few rare specimens being hydrothermally altered. Bennu represents one of these very rare types.

Most achondrites come from asteroids that were once molten. They underwent partial or complete separation into a metallic core, rocky mantle, and planetary crust, just like the Earth. Most achondrites are samples of the crust and mantle. Iron meteorites, composed primarily of iron nickel alloys, are mainly from the cores.

Spectral fingerprints make it possible to link meteorites with the asteroid types from which they originated:

- Carbonaceous chondrites are spectrally linked to the B- and C-type asteroids. This link was confirmed by Hayabusa2 samples from Ryugu.
- An exception to the above is the Tagish Lake carbonaceous chondrite, the only promising match for D-type asteroids.
- Ordinary chondrites are linked to the S-type asteroids. Hayabusa samples from Itokawa confirmed this link.
- Enstatite chondrites are tentatively linked to E-type asteroids.
- Iron meteorites are tentatively linked to the cores of M-type asteroids.
- Certain achondrites are linked to the crusts and mantles of V-type asteroids.
- Samples returned by Apollo astronauts confirmed that the lunar meteorites originated from the Moon.
- Measurements of the Martian atmosphere by NASA's Viking landers matched trapped gasses in the Martian meteorites, confirming their origin from the red planet.

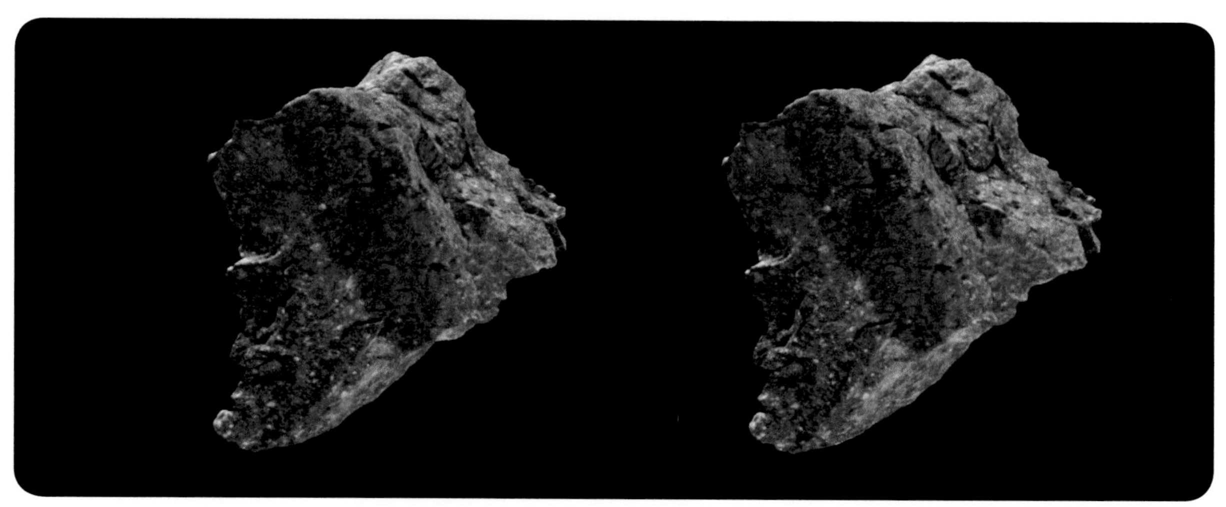

Two views of the meteorite Lonewolf Nunataks 94102, a carbonaceous chondrite from Antarctica that is just under 4 cm (1.4 in) in longest dimension. Bennu was predicted from remote sensing data to be made of similar material to meteorites of this type (Chapter 3). OSIRIS-REx scientists therefore studied it in preparation for the Bennu sample analysis campaign.

Intense hydrothermal alteration on the parent asteroids of some carbonaceous chondrites locked water into the minerals' crystal structures. These "hydrated" meteorites are also rich in carbon and carry a wide range of organic compounds. The hydrated NEAs from which they ostensibly came are the best targets for understanding how organic carbon and water made their way to the early Earth. The OSIRIS-REx mission embarked on a seven-year journey to Bennu and back to acquire a pristine sample of hydrated NEA material. Scientists will analyze this sample exhaustively, with the hope of shining a light on the origin of life.

The Atlas V rocket ready to launch OSIRIS-REx from Space Launch Complex 41 at Cape Canaveral Air Force Station, Florida.

Chapter 2

OSIRIS-REx, Asteroid Explorer

At 7:05 p.m. on September 8, 2016, a cloudless Florida evening, NASA launched its first asteroid sample return mission, the Origins, Spectral Interpretation, Resource Identification, and Security–Regolith Explorer — better known as OSIRIS-REx. The spacecraft shot into space aboard a United Launch Alliance Atlas V 411 rocket, initiating its seven-year expedition to rendezvous with asteroid Bennu in 2018, collect a sample in 2020, and bring it back to Earth in 2023.

OSIRIS-REx belongs to NASA's New Frontiers program of space missions that explore the Solar System to answer the highest-payoff questions in planetary science. It was selected in 2011 as the third mission in this program, following the Juno mission to Jupiter and the New Horizons mission to Pluto and the Kuiper Belt.

NASA's Goddard Space Flight Center in Maryland provided the overall management, systems engineering, and safety and mission assurance for OSIRIS-REx. Together with KinetX Aerospace, Goddard was responsible for navigating the spacecraft. Lockheed Martin Space built the spacecraft and handled flight operations. The University of Arizona led the science observation planning and data processing and was the headquarters of the mission's scientific team — which included researchers across North America, Europe, Asia, and Australia.

The extraterrestrial nature of the sample delivered to Earth by OSIRIS-REx requires a special curation facility, which is housed at NASA's Johnson Space Center in Houston, Texas. One-quarter of the mass of the sample is allocated to the OSIRIS-REx science team for immediate analysis. The Canadian Space Agency receives 4% of the sample in exchange for their contribution of the onboard laser altimeter (described later in this chapter). The Japan Aerospace Exploration Agency (JAXA) receives 0.5% as part of an ongoing interagency collaboration. The remaining 70.5% will be set aside for analysis by the global research community, now and well into the future.

The rocket carrying the OSIRIS-REx spacecraft launches from Cape Canaveral, on September 8, 2016.

View of the OSIRIS-REx spacecraft from a camera onboard the rocket as it detaches for the journey to Bennu. The spacecraft is wrapped in thick thermal blanketing (germanium kapton mylar, which resembles aluminum foil in this image) to protect it from the space environment and keep its components at operating temperatures.

Why sample return?

The OSIRIS-REx mission could have been greatly simplified by observing Bennu without sampling it. What made the operational complexity, risk to the spacecraft, and cost of sample collection worthwhile? Observing an extraterrestrial object from close range has immense scientific value, but without collecting a sample, the gains in knowledge are limited by the design and capabilities of the spacecraft. With sample return missions such as the Apollo program at the Moon and OSIRIS-REx at Bennu, samples can be archived for future generations, who will pose questions and use technologies that have not yet been conceived.

Although meteorites are naturally delivered samples (Chapter 1), only the sturdiest survive passage through the Earth's atmosphere. This natural filter biases our sample archive, which therefore does not accurately represent our small Solar System neighbors. Moreover, until they are found and collected, meteorites are exposed to air, water, soil, and biological organisms that can confuse the interpretation of their chemical signatures. Though we can infer the likely history of a meteorite, we cannot be certain of the specific place from which it originated. Collecting a sample from a carefully observed extraterrestrial location such as Bennu, and preserving it through atmospheric entry and landing, maximizes what we can learn from it.

OSIRIS-REx is not the only modern mission to recognize this value. In parallel with the OSIRIS-REx encounter with Bennu, JAXA sent the Hayabusa2 mission to visit and collect a sample from the asteroid Ryugu. Bennu and Ryugu resemble each other, with some intriguing physical and spectral differences, suggesting a similar origin followed by different geologic histories. The partnership between these two missions increases the scientific return for both, offering opportunities to compare data and learn from one another.

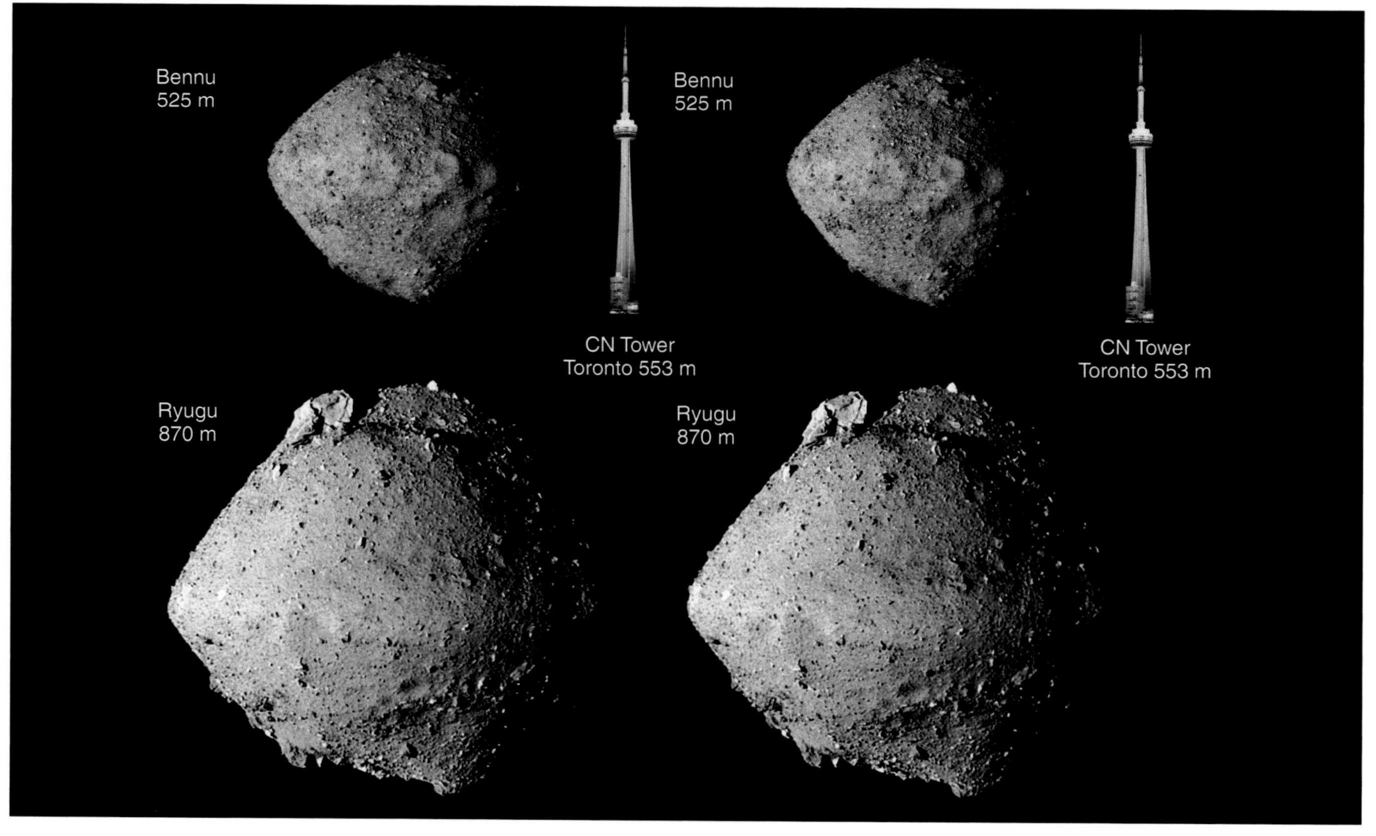

These stereos show how Ryugu resembles Bennu, on a larger scale. Both are diamond-shaped spinning piles of rubble.

Why Bennu?

The target of the OSIRIS-REx mission was selected by the team in 2005. Of the hundreds of thousands of asteroids that had been identified in our Solar System at that time, what drove OSIRIS-REx to visit Bennu in particular?

The first considerations were based on spacecraft engineering constraints. Near-Earth asteroids presented an obvious advantage over those farther afield in terms of the timeframe and resources needed to reach the target and return to Earth. Among the thousands of near-Earth asteroids that were known at the time of target selection, 192 had orbits that were accessible within the limits of spacecraft design.

Of these reachable objects, 26 had diameters larger than 200 meters (219 yards). This size threshold was desirable because spacecraft operations can be undesirably complicated around smaller asteroids. They can have very fast spin rates — rotating once every two or fewer hours — and non-principal axis rotation, where the asteroid tumbles or wobbles. Further, such rapid rotators have large centrifugal forces that can fling loose regolith into space. Thus, they might not have the type of surface material that OSIRIS-REx was designed to collect.

Within this subset of 26 feasible targets, relative scientific value could finally come into play. Five of the 26 asteroids had spectral signatures consistent with being carbon-rich. Such carbonaceous asteroids are probably relics of the early Solar System (Chapter 1) and contain chemical compounds that were crucial to the formation of Earth as a habitable planet, such as water and organic molecules. Though rare in the near-Earth population, carbonaceous asteroids, also known as C-types, constitute the most populous asteroid class in the main belt.

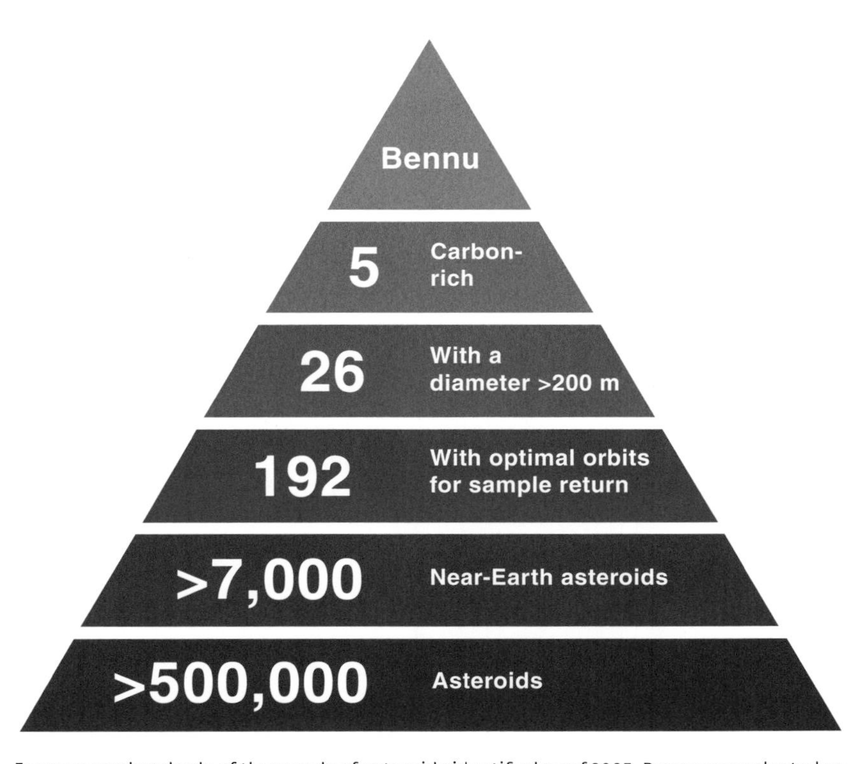

From among hundreds of thousands of asteroids identified as of 2005, Bennu was selected as the mission target on the basis of proximity to Earth, favorable orbital dynamics for sample return, sufficient size for operational feasibility, and carbonaceous composition.

Of the five carbonaceous options, one — Bennu — has an orbit that brought it close enough to Earth to enable extensive observation by ground- and space-based telescopes (Chapter 3) in the years leading up to the proposed mission. The knowledge of Bennu's properties and behavior gained during these close encounters with Earth ultimately informed the design of the spacecraft operations and contributed to the mission's success.

The OSIRIS-REx flight system

The OSIRIS-REx flight system consists of the spacecraft, the machinery for collecting the sample, and six science instruments for characterizing Bennu. The spacecraft gets power from two wing-like solar panels, which make it just over six meters (20 feet) wide when deployed. Twenty-eight engines fed by hydrazine propellant are used for maneuvers (known as "burns") that change the spacecraft direction and velocity. A high-gain dish antenna enables Doppler tracking and communication with Earth.

The instruments aboard the spacecraft sense a much wider range of wavelengths than humans do, as illustrated in the diagram overleaf. By remotely sensing this diversity of wavelengths, OSIRIS-REx scientists were able to learn a substantial amount about Bennu before attempting the delicate sampling maneuver, ensuring its safety and success.

The OSIRIS-REx flight system during final processing at the Kennedy Space Center, six days before launch.

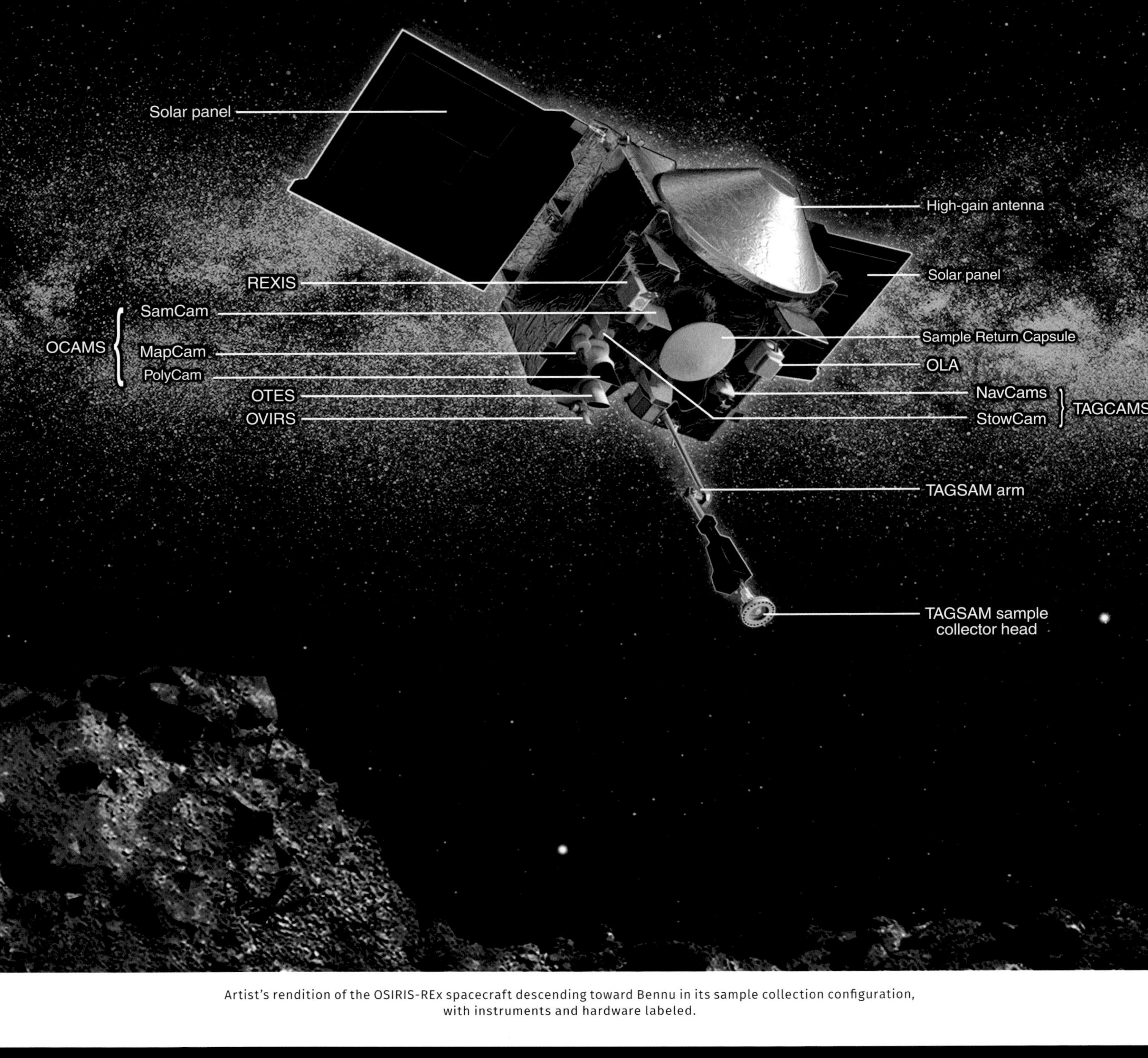

Artist's rendition of the OSIRIS-REx spacecraft descending toward Bennu in its sample collection configuration, with instruments and hardware labeled.

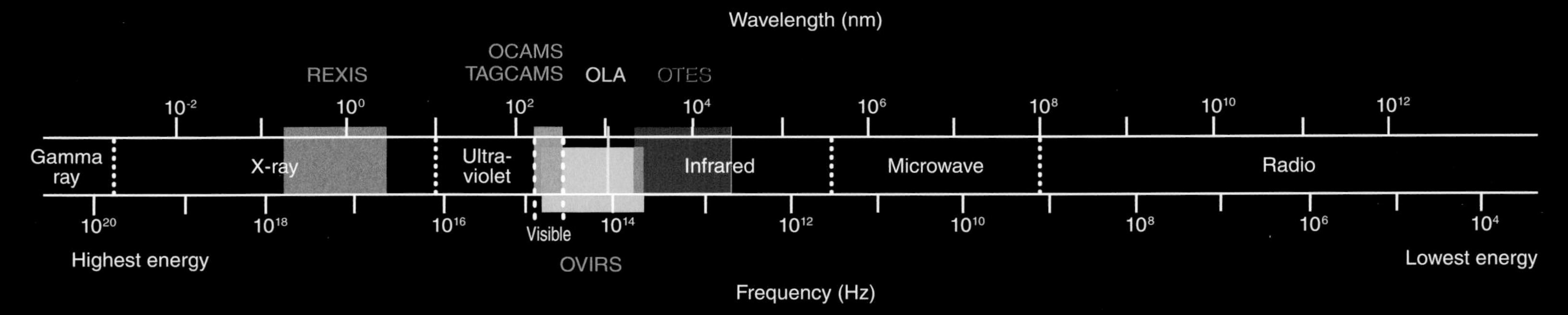

Wavelengths observed by the OSIRIS-REx science instruments.

Items onboard the flight system:

- The **Regolith X-Ray Imaging Spectrometer (REXIS)**, a student-run instrument, was designed to detect fluorescent X-rays indicating the abundances of various elements on Bennu. Surprisingly, Bennu apparently does not emit X-rays, a scientific puzzle yet to be solved. REXIS also detected a newly flaring black hole in the constellation Columba.

- The **OSIRIS-REx Camera Suite (OCAMS)** imaged the asteroid at global to local scales, providing data to construct the maps and digital terrain models needed for effective and safe sampling. OCAMS includes:

  the close-range **SamCam**, which documented the sample collection event;

  the multispectral **MapCam**, which used four color filters — blue, green, red, and near-infrared — to map the diversity of surface materials;

  and the telescopic **PolyCam**, the only camera on board with a zoom lens, which first detected Bennu from more than two million kilometers (1.24 million miles) away and ultimately revealed its surface in millimeter-scale detail.

- The **OSIRIS-REx Laser Altimeter (OLA)** fired laser pulses at Bennu's surface to measure its distance from the spacecraft. These ranging data were used to construct exquisitely detailed digital terrain models of the asteroid's shape and topography, complementing those produced from OCAMS images and providing highly accurate maps of hazards around the candidate sample sites.

- The **Touch-and-Go Camera System (TAGCAMS)**, another camera suite, in this case designed to support operations, though it also supported science serendipitously (Chapter 3). TAGCAMS includes:

  two identical **NavCams**, one prime and one backup, which imaged Bennu's surface, the near-Bennu environment, and the background starfield, to enable spacecraft navigation, including during the descent to the surface for sample collection;

  and the **StowCam**, which documented the safe stowage of the sample in the return capsule.

- The **OSIRIS-REx Visible and near-InfraRed Spectrometer (OVIRS)** collected spectra for mapping the organic, mineral, and water content of Bennu's surface materials. These spectral datasets informed the selection of a sample site with the desired composition.

- The **OSIRIS-REx Thermal Emission Spectrometer (OTES)** detected longer-wavelength infrared spectra that were used to map minerals and temperature, as well as to model thermal inertia. Interpretation of these data ensured that the surface would not be dangerously hot for the spacecraft to contact.

- The **Touch-and-Go Sample Acquisition Mechanism (TAGSAM)**, a pogo stick-like instrument consisting of a robotic arm and sample collector head, unfolded from the spacecraft during the sample collection maneuver. TAGSAM fired high-pressure nitrogen gas into Bennu's surface to propel loose rocky material into the head. The first sampling attempt was successful, but TAGSAM had enough gas for two more tries, had they been needed.

- The **Sample Return Capsule (SRC)** encased the TAGSAM head containing the collected sample to protect it during the final maneuvers at Bennu and the inbound cruise to Earth. This capsule — the only component of the spacecraft destined to return to Earth's surface — was equipped with a heat shield and parachutes to help it withstand atmospheric reentry and land safely.

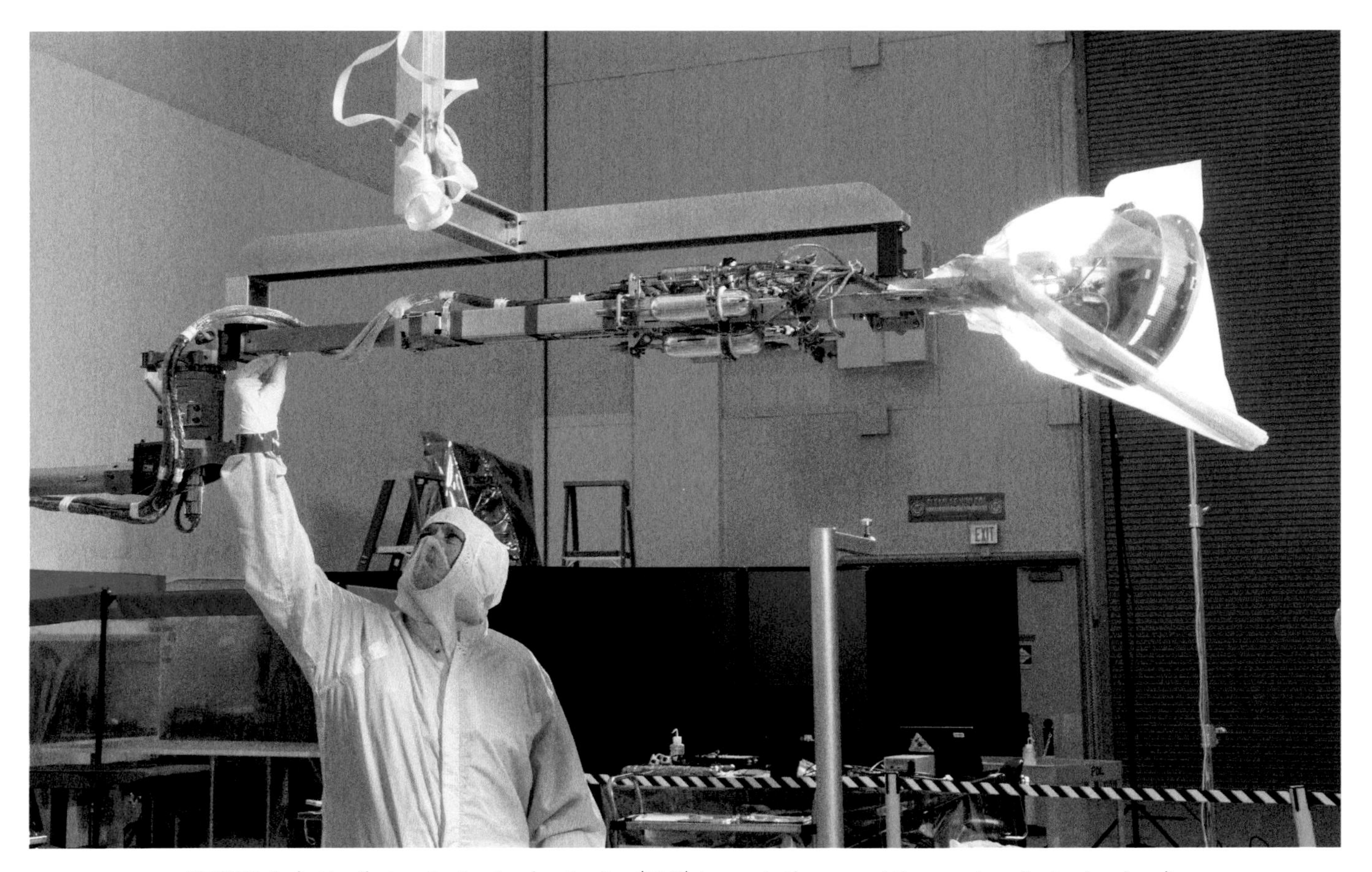

TAGSAM during testing on Earth, showing the 3 m (10 ft)-long robotic arm and the sample collector head on its end. Here, the head is enclosed in plastic to prevent contamination and ensure a pristine sample. Three bottles containing high-pressure nitrogen are visible near the end of the arm. Each bottle contains enough gas for one sampling event, affording the mission three chances to collect a sample.

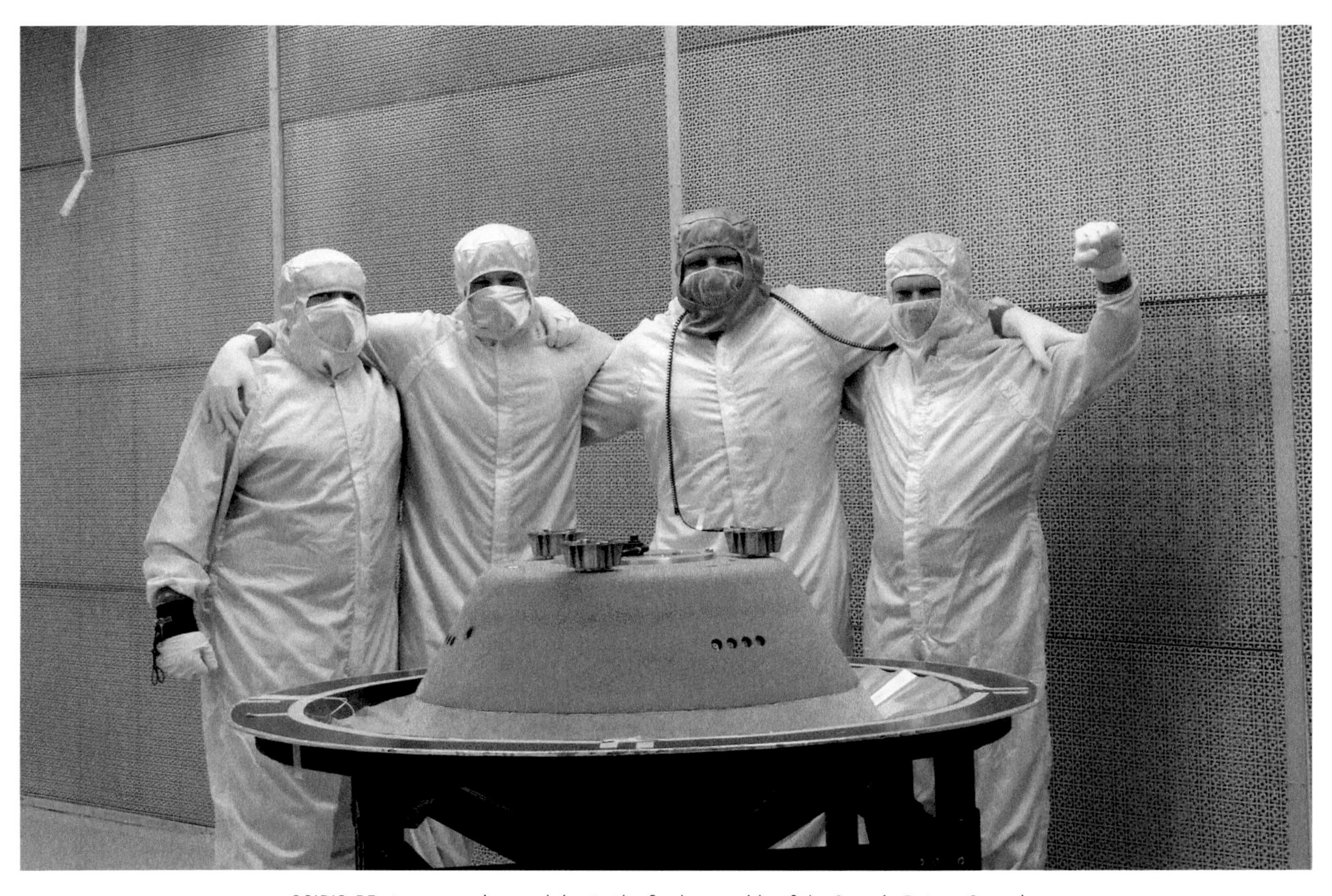

OSIRIS-REx team members celebrate the final assembly of the Sample Return Capsule.

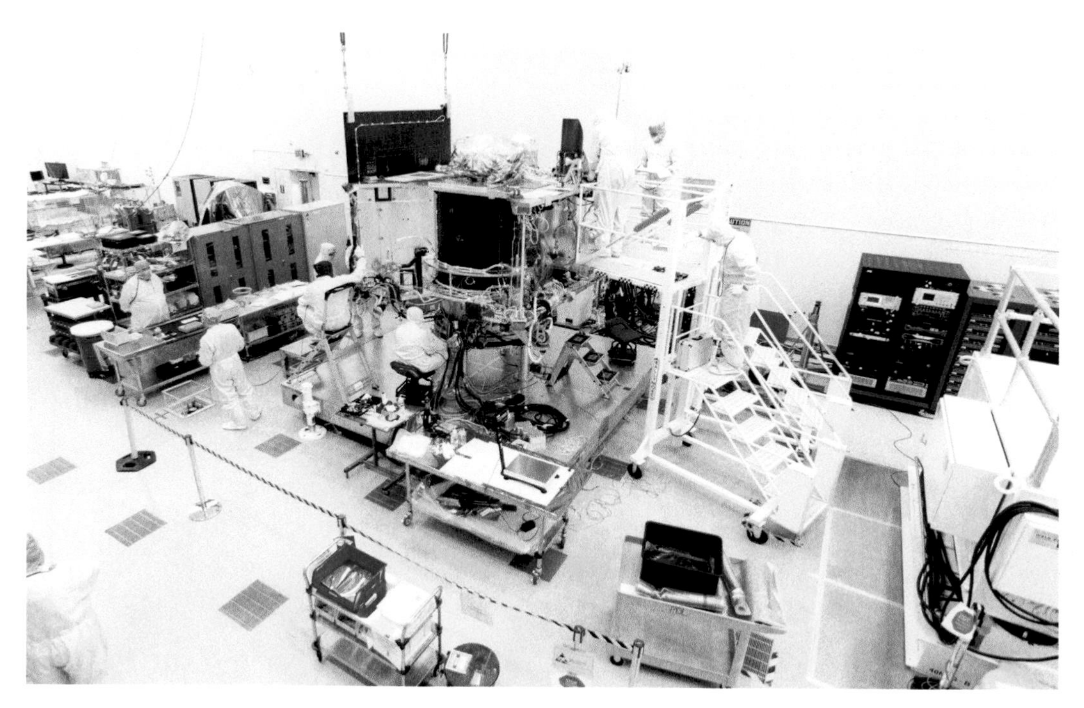

Installing OCAMS on the spacecraft.

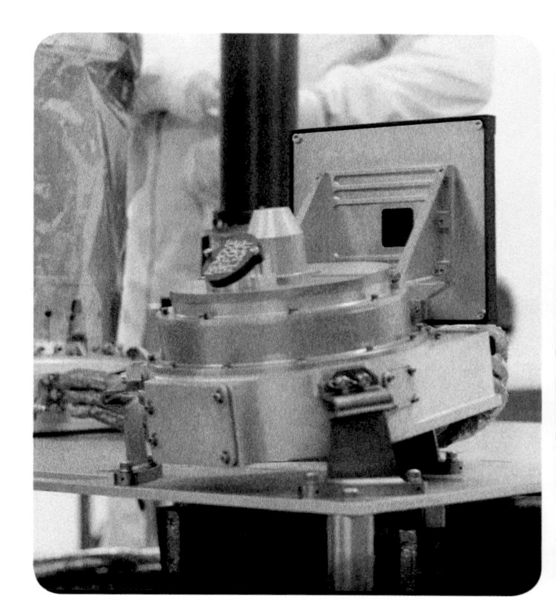

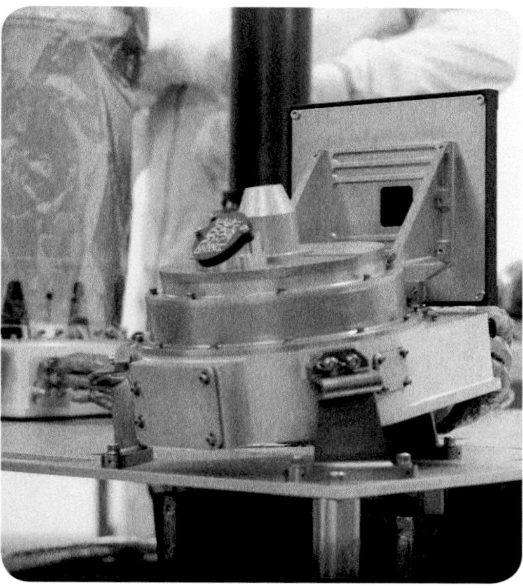

Stereo views of the OCAMS imagers in the laboratory. MapCam (small, left) and PolyCam are shown in the upper pair, and SamCam is shown in the lower pair.

The outbound journey

The rocket that launched OSIRIS-REx off the surface of Earth in 2016 provided enough momentum to propel the spacecraft towards Bennu. However, the asteroid's orbit around the Sun is tilted six degrees from Earth's, and the rocket did not have enough energy to match this angle. OSIRIS-REx needed an extra boost from Earth's gravity to change its orbital plane. This meant that the spacecraft had to travel all the way around the Sun to meet Earth one year after launch. On September 22, 2017, the spacecraft used Earth's gravity to slingshot itself to rendezvous with Bennu.

At closest approach, the spacecraft came within 17,237 kilometers (10,711 miles) of Antarctica, just south of Cape Horn, Chile, before heading north over the Pacific Ocean. This Earth gravity assist changed the angle of the spacecraft's orbital plane, putting it on a direct path toward Bennu, and changed the spacecraft's velocity by about four kilometers (2.5 miles) per second.

In addition to the gravity assist, the Earth flyby provided the first opportunity to check out the onboard instrument suite. For two weeks, the science instruments scanned Earth and the Moon. This opportunity to collect science data allowed the OSIRIS-REx team to practice for operations at Bennu.

This color image of Earth was taken on September 22, 2017, by MapCam just hours after the spacecraft completed its Earth gravity assist, at approximately 170,000 km (106,000 mi) away from the planet. The Pacific Ocean and several continental landmasses are visible. Australia is in the lower left; Baja California and the southwestern United States appear in the upper right. (The black lines at the top are imaging artifacts.)

On September 25, 2017, PolyCam imaged the Moon from a range of 1.2 million km (0.75 million mi). Familiar lunar nearside features such as the Sea of Tranquility, site of the Apollo 11 landing, and the Sea of Crises, target of the Soviet lunar sample-return mission Luna 24, are visible on the left. Most of the image is of the far side of the Moon, which is never visible from the surface of the Earth. To produce this image, the mission team combined nine images using a technique called super-resolution imaging.

This color image of Earth and the Moon was acquired with the MapCam on October 2, 2017. The distance to Earth was approximately 5,120,000 km (3,181,420 mi) — about 13 times the distance between Earth and the Moon.

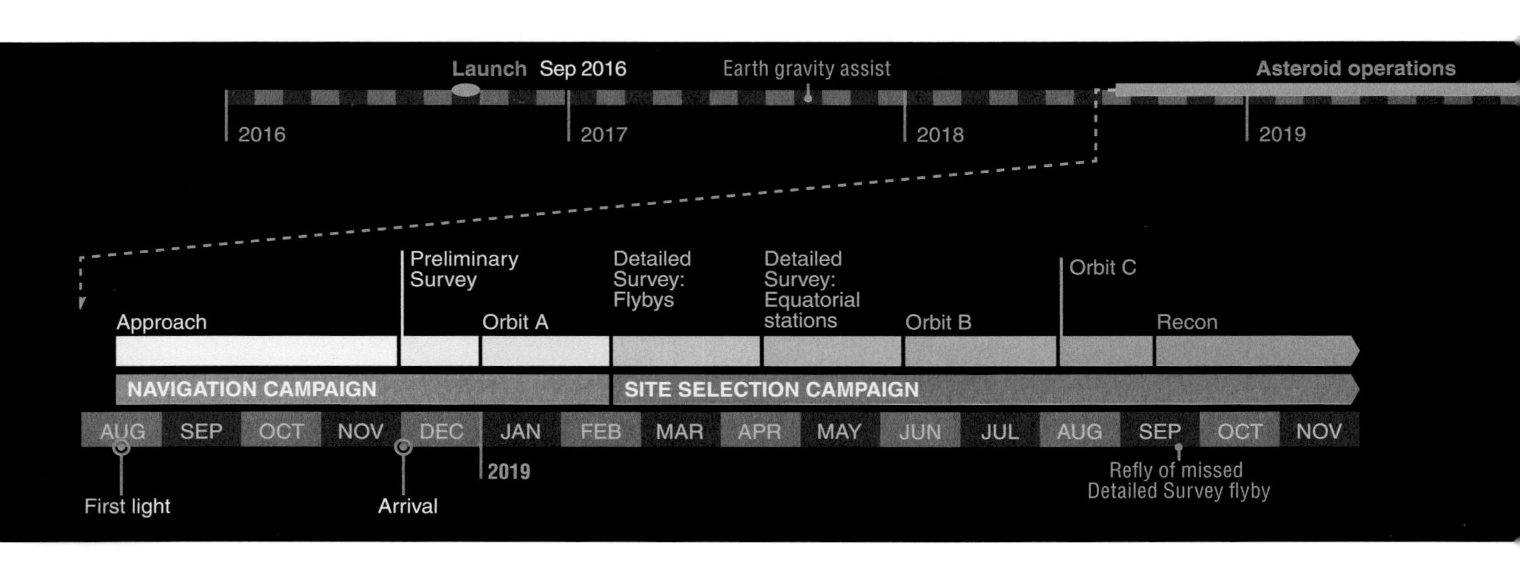

Up close with Bennu

Almost 11 months after the encounter with Earth, on August 17, 2018, PolyCam caught its first glimpse of Bennu from a distance of more than two million kilometers (1.24 million miles). This "first light" marked the beginning of the mission's Approach phase. The primary objectives of this phase were to survey the surrounding area for potential hazards such as moonlets and dust plumes. As Bennu transformed from a point of light into a resolved world, the team collected enough images to generate a shape model of the asteroid, assign a coordinate system (Chapter 7), and measure its rotation rate.

Over the next 16 months, the mission progressed through a series of phases, each designed to collect specific information needed to select the best location for sampling Bennu.

Over the poles and along the equator

On December 3, 2018, the spacecraft fired a miniscule blast from its rocket engines, consuming just a few grams of propellant. This maneuver changed the trajectory from approaching Bennu to scrutinizing the asteroid. The event signified the formal arrival at Bennu and marked the beginning of the Preliminary Survey phase.

For this initial survey, the team needed to observe Bennu from three different viewpoints: over the north pole, along the equator, and under the south pole. The closest approach for each polar or equatorial pass was seven kilometers (4.35 miles).

One of the greatest challenges in arriving at Bennu was that the precise position of the spacecraft relative to the asteroid was not easy to determine. On December 3 this uncertainty was so large that the team decided to execute a triple pass of the north pole, zipping back and forth three times to refine knowledge of the asteroid's position and its microgravity environment. This strategy enabled exact positioning for the best science observations. The third pass put the science instruments on the right path to get the best possible data.

Now, with the spacecraft's positional uncertainty reduced, the equatorial and southern passes only needed to be performed once. The spacecraft made sharp, precise turns

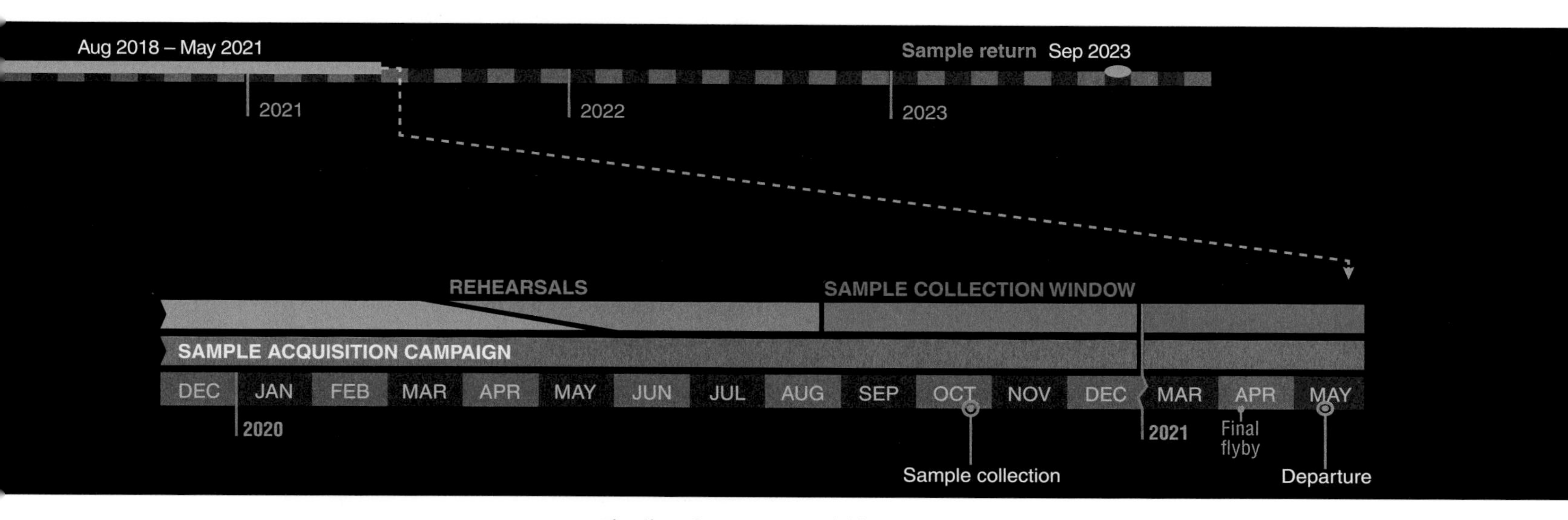

Timeline of OSIRIS-REx activities at Bennu.

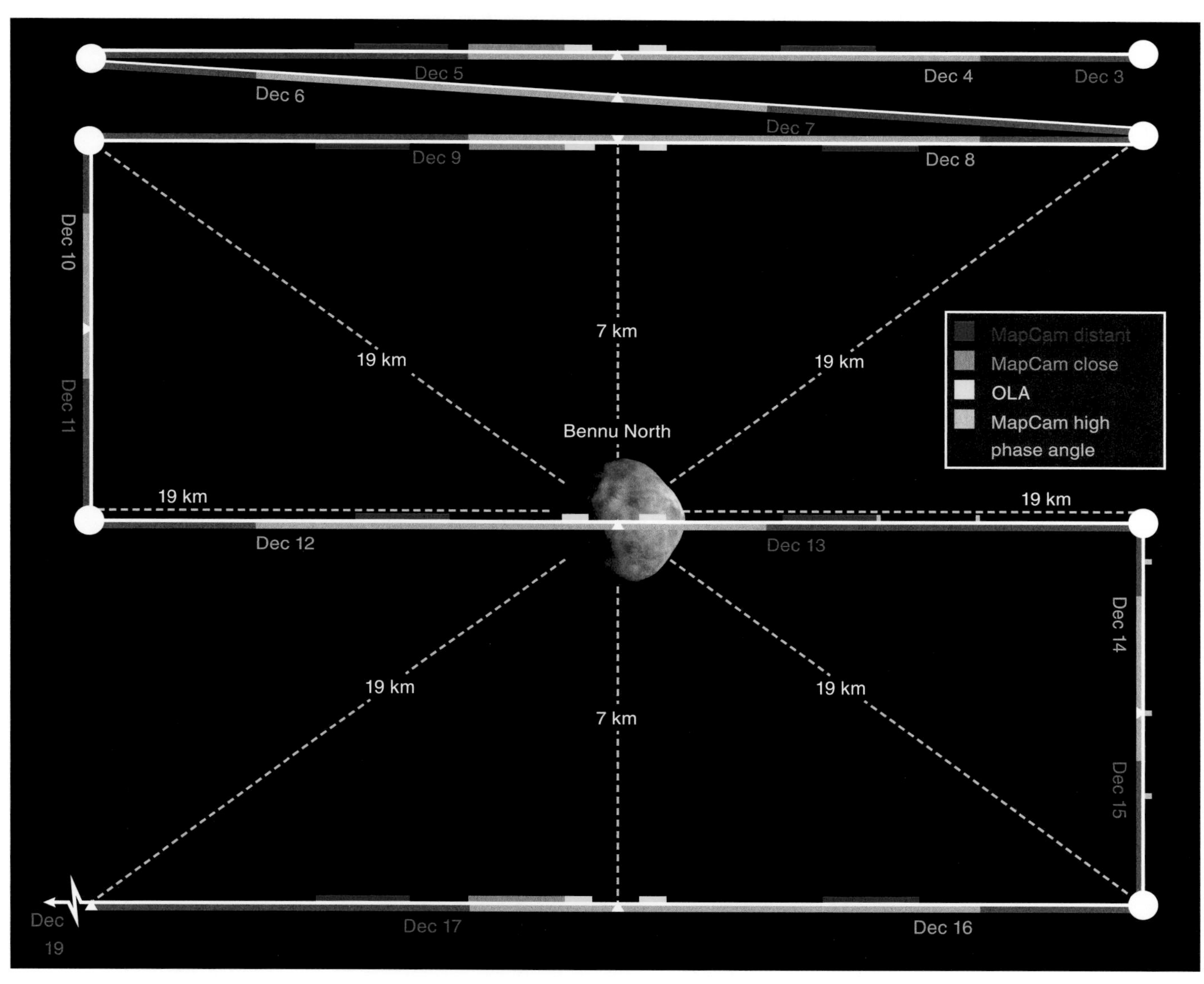

The spacecraft's trajectory around Bennu during the Preliminary Survey in 2018, as viewed in the plane of the equator.

to transition southward along the planned path. Carefully monitoring Bennu's slight gravitational tug on the spacecraft at each closest approach made it possible to measure the asteroid's mass. The images collected on these passes refined the team's knowledge of Bennu's rotation and improved the resolution of the global shape model.

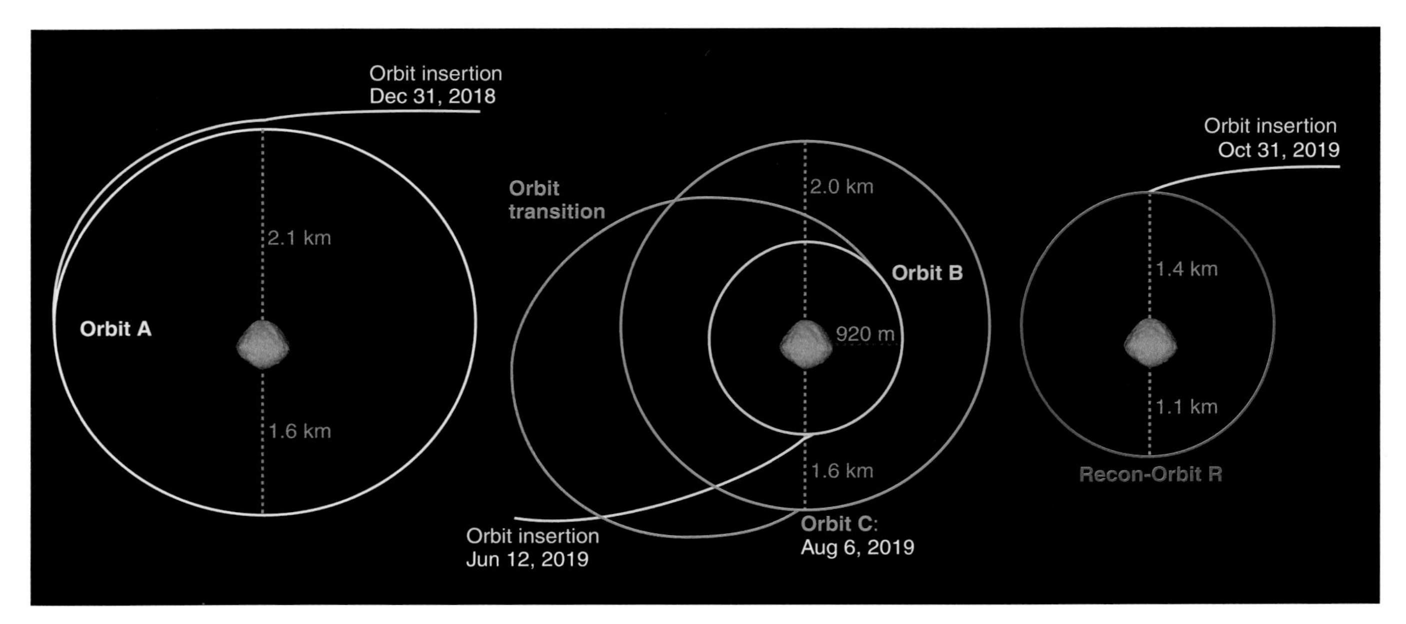

The 4 orbital campaigns around Bennu.

In and out of orbit

While the scientists were taking their first detailed looks at the asteroid's surface, the mission's navigation team used the Preliminary Survey to practice the delicate art of navigating in microgravity. On December 31, 2018, the team rang in the New Year by entering orbit, making Bennu the smallest object ever to be orbited by a spacecraft. The spacecraft circled Bennu at a distance between 1.6 and 2.1 kilometers (1 and 1.3 miles), traveling at the sluggish velocity of five centimeters (two inches) per second, making each orbit last just over 61 hours. This feat marked another milestone in spaceflight: This range was the closest that a spacecraft had ever orbited around a celestial body.

During this first orbital phase (Orbit A), the navigation team transitioned from star-based navigation to landmark-based navigation. Using landmarks such as rocks and craters on Bennu's surface, they were able to determine the spacecraft's position precisely enough for the detailed observations in the upcoming mission phases.

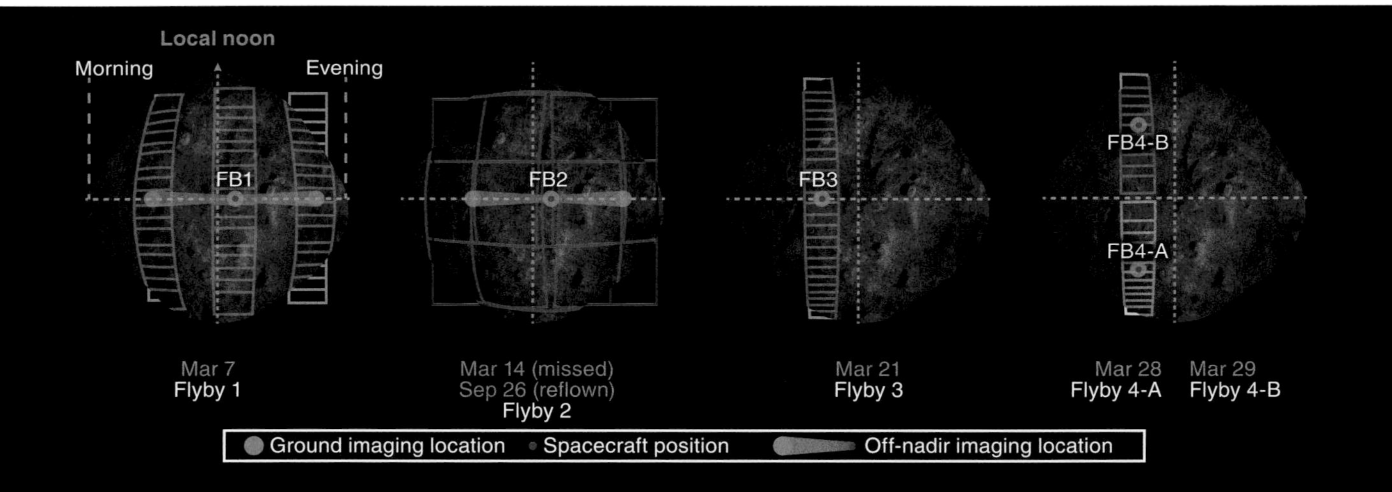

Flybys 1 to 7 of the Detailed Survey in 2019. The original Flyby 2 was missed owing to a severe storm in Colorado that caused a communications outage, which prevented the team from uplinking the final commands to the spacecraft.

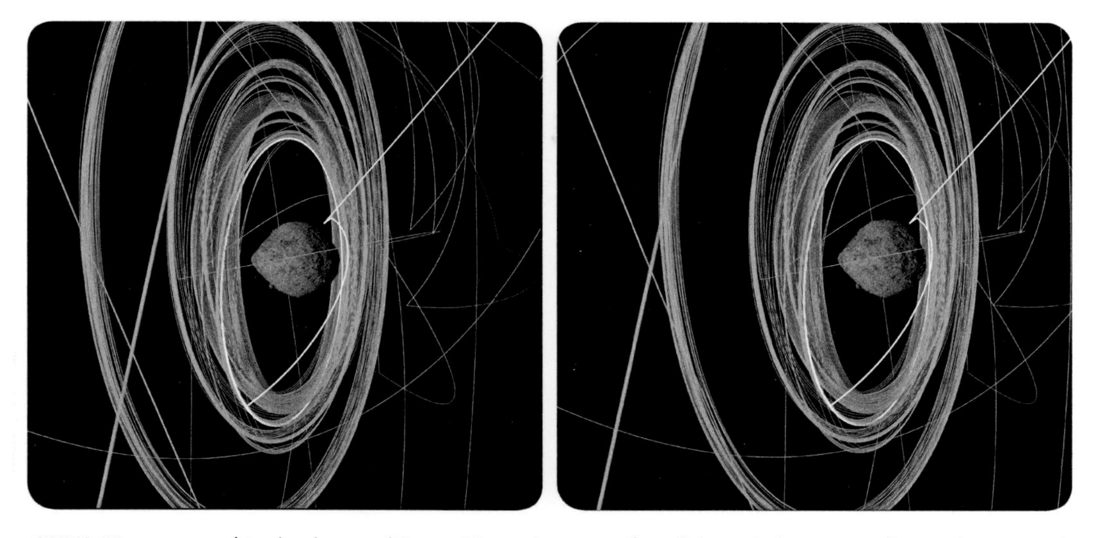

OSIRIS-REx weaves a virtual web around Bennu. We see here a portion of the path the spacecraft traced out around the asteroid, up to the completion of the first sampling rehearsal. The brighter part of the track is the most recent.

Being in orbit around Bennu was the most stable configuration possible for the spacecraft, allowing the team to predict its future location with high confidence. The downside of this positioning was that the imaging conditions were poor. The spacecraft was always looking at dawn or dusk on Bennu, resulting in long shadows and dim illumination. To collect more detailed global data, the spacecraft would need to leave orbit and get into a position closer to noon, local solar time. These campaigns were excursions. After each one, the spacecraft returned to its "safe home" in orbit around Bennu.

Of the four orbital campaigns, only Orbit B was dedicated to science. For this phase, the spacecraft maintained a constant, very close distance of 920 meters (1,006 yards) from Bennu. This consistency was needed to allow OLA to collect data on the roughness and topography of the asteroid. The later orbit phases (designated C, for the third orbit, and R, for the Reconnaissance phase) were waypoints between the more demanding observational campaigns.

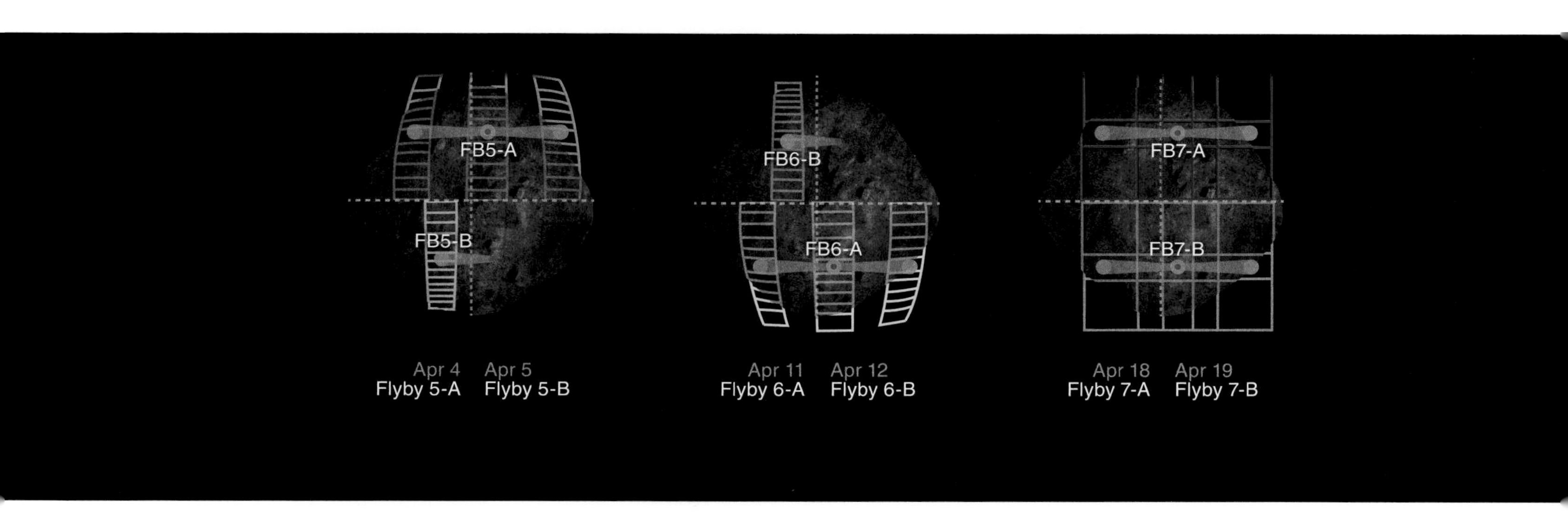

Let the mapping begin

The in-depth study of Bennu began in earnest during Detailed Survey, which kicked off on February 28, 2019. During this phase, OSIRIS-REx made multiple passes over Bennu to obtain the wide range of viewing angles necessary to capture details of the geology, topography, composition, and temperature over the entire global surface. The team spent most of 2019 analyzing these data to identify candidate sample sites.

Detailed Survey consisted of two seven-week campaigns in which the spacecraft observed Bennu once per week. The first seven weeks were devoted to Flybys 1 through 7, at a range of just over three kilometers (1.9 miles). In Flybys 1 to 3, the spacecraft observed from a position close to the equator; in Flybys 4 to 7, it observed from positions to the north and south. Some images were obtained close to local noon, when there were almost no shadows, so that variations in the brightness of surface materials could be discerned. Others were captured with the spacecraft looking toward the morning or evening on Bennu, so that the long shadows would reveal the rugged topography of the asteroid. PolyCam provided the best resolution, while MapCam captured color images.

In the second seven-week campaign, the spacecraft remained near the equator at a range of about five kilometers (3.1 miles) and observed Bennu at seven different local times. These data were critical for learning how Bennu's surface heats up and cools off during each rotation and for acquiring the best-quality compositional data from the spectrometers.

At the end of these grueling 14 weeks, the team rested knowing they had performed the most comprehensive asteroid mapping endeavor in history.

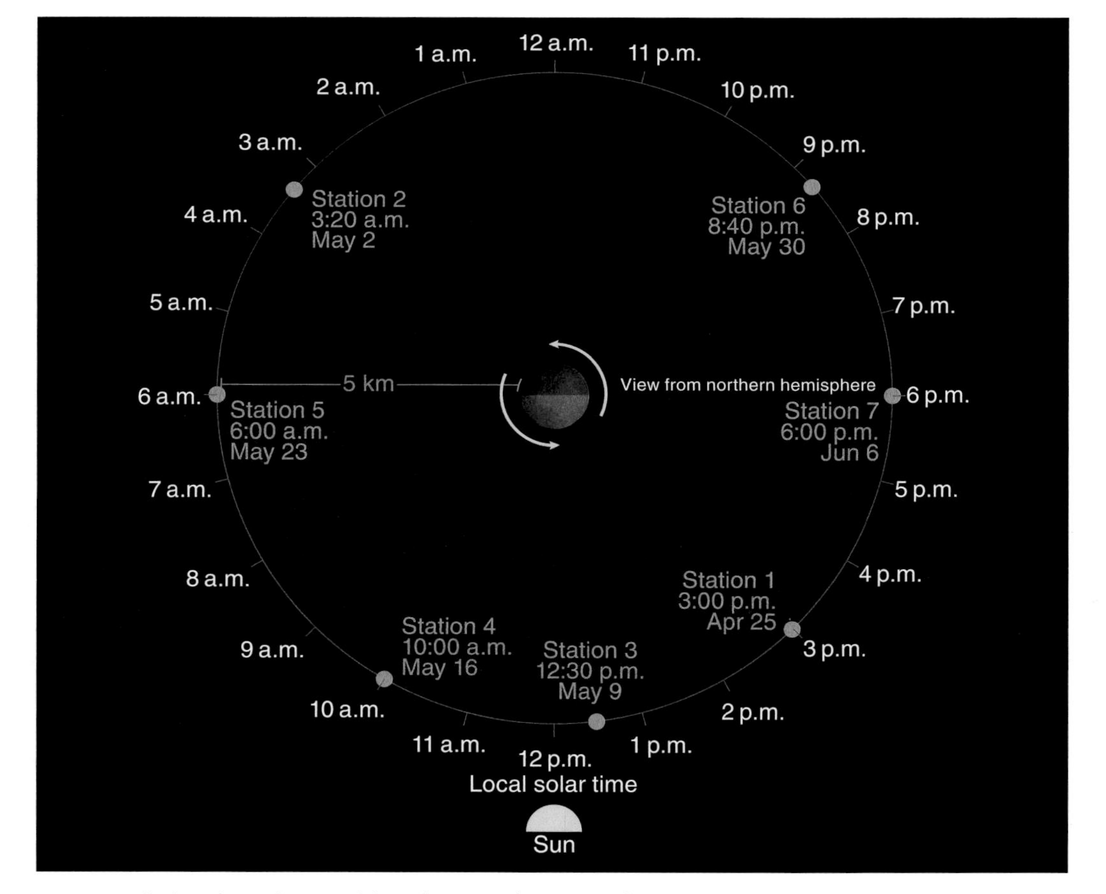

The locations of Equatorial Stations 1 to 7 in 2019, as viewed from above Bennu's north pole.

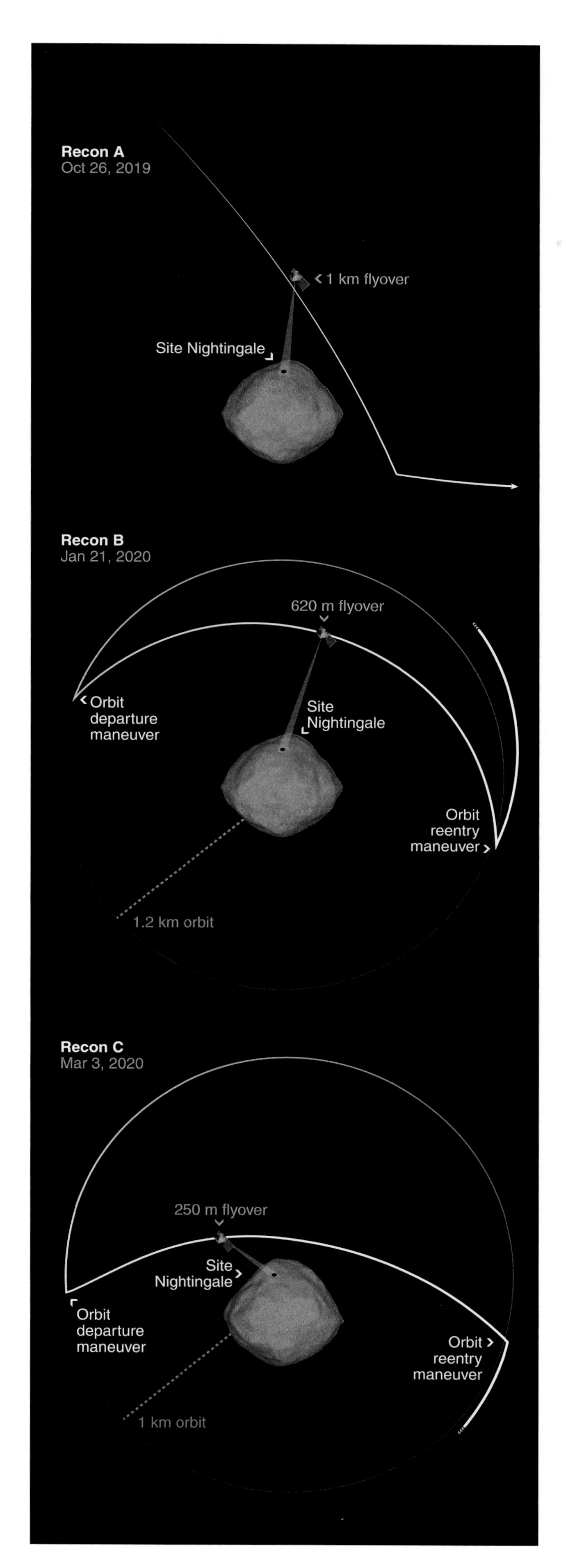

Reconnaissance expeditions

The data collected in Detailed Survey and Orbit B provided the team with enough information to narrow down the number of potential sample sites on Bennu to the four best candidates (Chapter 5). The next step was to get closer to the asteroid for a better look at these areas.

During the Reconnaissance (Recon) A phase, in October 2019, the spacecraft performed four flyovers — one for each candidate site — at a distance of about one kilometer (0.6 miles) from Bennu. Each site was observed along a sweeping trajectory that began days before the closest approach. The team had to time each pass very carefully so that the area of interest was directly below the spacecraft, with the correct lighting conditions and observing angles, when data collection began. After each pass, the spacecraft took a sharp turn to reset for the following week's viewing.

The spacecraft then retreated to a safe-home orbit, Orbit R (for "Reconnaissance"), as the team poured over the data. By Recon B, in early 2020, the team had narrowed the sites down to two: a primary site that would be prioritized for the first sampling attempt, and a backup site in case that attempt failed. For the flyovers of these two sites, OSIRIS-REx came even closer, to a range of 625 meters (684 yards) from the surface. These passes collected data to inform the spacecraft's autonomous guidance system, which was needed for a precision touchdown.

Recon C, in spring 2020, was the last chance to check the suitability of the two sites. These final two sorties approached Bennu to a mere 250 meters (820 feet), allowing the cameras to image the surface at scales of a few millimeters. These images were crucial for identifying the best point within each site to target for sample collection.

The trajectories of the three Reconnaissance campaigns over the primary sample site, nicknamed Nightingale (Chapter 5). In Recon B and C, the spacecraft stayed in a safe-home orbit (Orbit R) in between flyovers.

Time to play TAG

Because sample collection was a highly demanding, mission-critical event, the team rehearsed it twice, each time stopping short of contact. During the first rehearsal, in April 2020, the spacecraft left orbit, maneuvered to a predefined Checkpoint 125 meters (410 feet) above the primary site, and then returned to orbit. The second rehearsal, in August 2020, took the spacecraft from orbit, through the Checkpoint, and down to the Matchpoint, further testing the ability of the onboard autonomous guidance system to correct its trajectory in real time. Everything was now prepared for the Touch-and-Go (TAG) sample collection event.

Finally, on October 20, 2020, OSIRIS-REx steered itself to the surface with the TAGSAM extended, and the momentum of the spacecraft pushed the sample collector head into the asteroid (Chapter 6). One second after contact, TAGSAM blew its nitrogen gas down into the subsurface to mobilize dust, pebbles, and gravel, directing them into the collection chamber inside the head. After stowing the sample in the return capsule, the spacecraft slowly drifted away from Bennu for several months.

TAG was meant to be the last encounter with Bennu. However, the event was so dynamic — sending up an unexpectedly massive plume of dust and debris (Chapter 6) — that the team made a last-minute decision to return to the asteroid for a final flyby on April 7, 2021, similar to Flyby 1 from the Detailed Survey. Having documented the disturbance to the surface caused by TAG, the team prepared for departure and delivery of the sample to Earth.

The following chapters document the extraordinary data set collected by this daredevil mission. Prepare to marvel at the beauty of asteroid Bennu, a small but significant member of our Solar System.

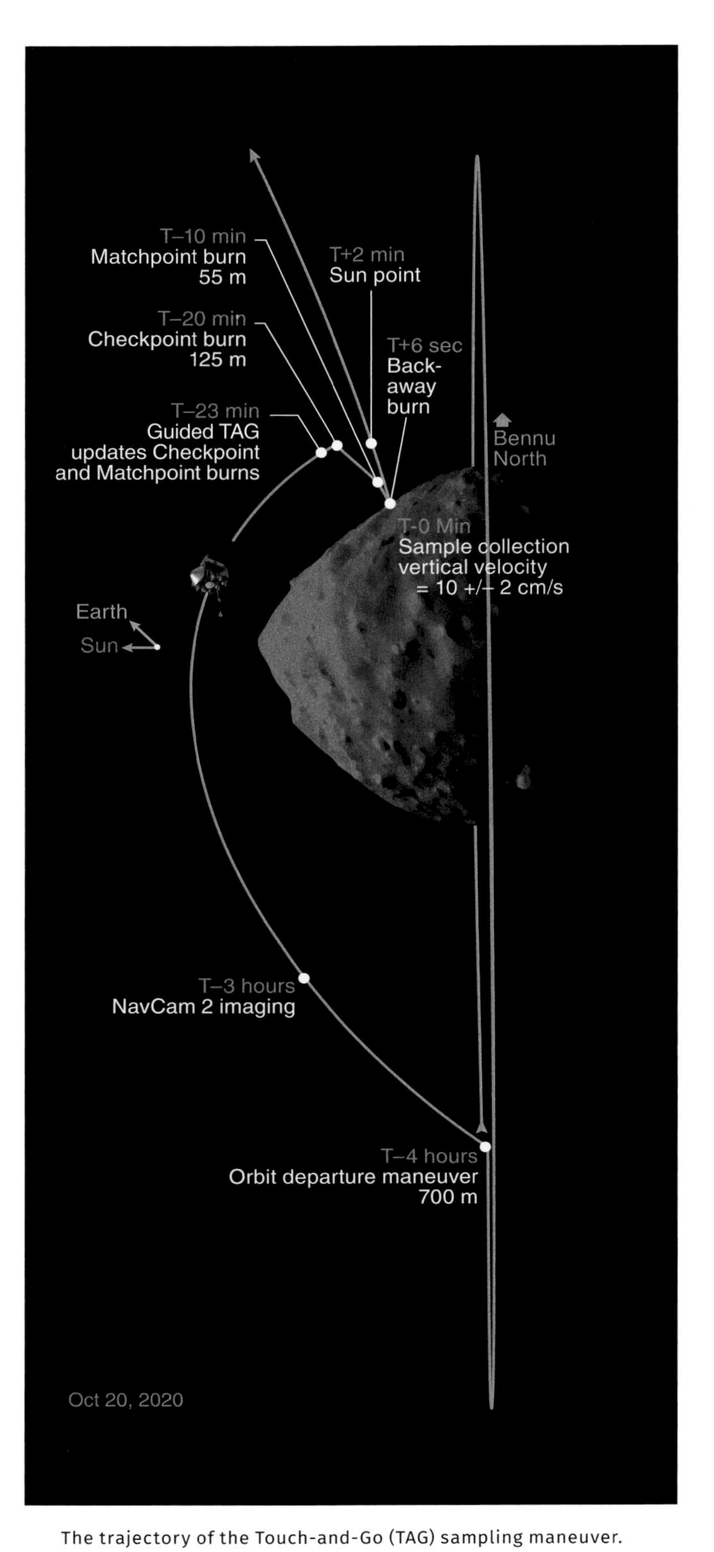

The trajectory of the Touch-and-Go (TAG) sampling maneuver.

During a routine checkout of the TAGSAM arm articulation on November 14, 2018, SamCam captured images with the TAGSAM head in the foreground and Mars as a tiny point of light in the distance. The image shown here is a composite of two SamCam images: one in which the TAGSAM head could be seen clearly but Mars was too dark, and one in which Mars was visible but the TAGSAM head was overexposed.

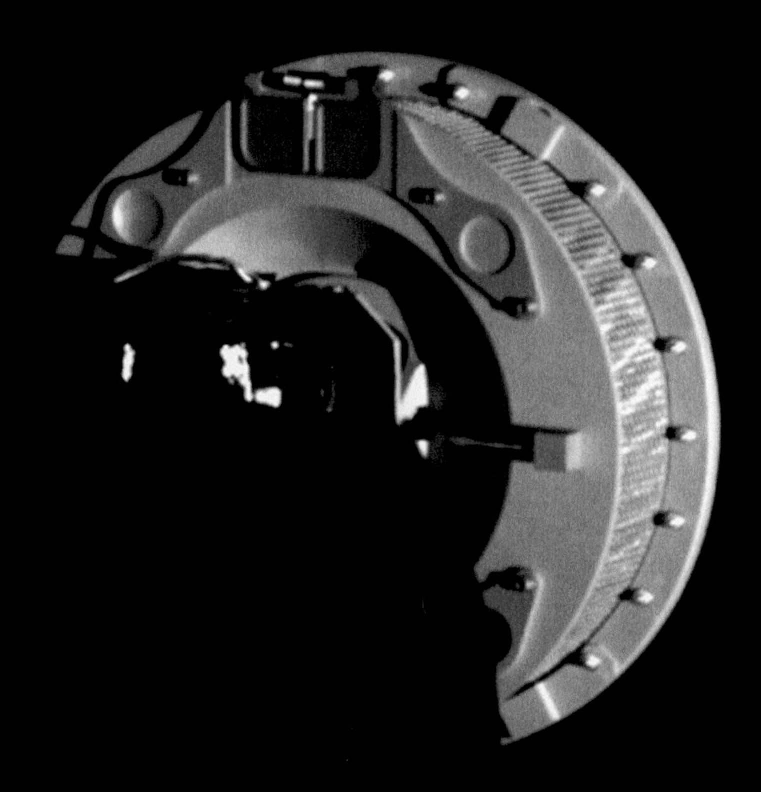

Bennu emerges from the void as the OSIRIS-REx spacecraft closes in. These PolyCam images were taken between late October and November 2018. The iconic boulder BenBen, about as tall as the Statue of Liberty, hangs off the southern hemisphere horizon.

Chapter 3

Welcome to Bennu

Asteroid Bennu is one of the oldest, darkest, and most intriguing objects in our Solar System. Its orbit is very similar to that of Earth, and it has a dark surface with a reflectivity similar to that of coal. In addition, Bennu is very small in planetary terms, only about half a kilometer (one-third of a mile) at its widest point.

Bennu's orbit draws it close to Earth every six years, offering astronomers regular opportunities for observation. Indeed, Bennu was discovered during one such close approach in 1999 by the Lincoln Near-Earth Asteroid Research (LINEAR) survey. As soon as the discovery was announced, observatories around the world locked on to Bennu's location to study this newly-identified celestial neighbor.

The Klet Observatory in the Czech Republic captured Bennu in a series of images during the discovery apparition. The two halves of this stereo pair were made using two 30-second-exposure images taken roughly 3 min apart on September 13, 1999. When viewed stereoscopically, Bennu can be seen in the foreground against a background of stars.

The scientific community observed Bennu extensively during close approaches in 1999, 2005, and 2011, when it was near enough to be seen with telescopes. The observations made possible by these apparitions were used to create the most detailed model, at the time, of an asteroid orbit — one that included reflected sunlight and emitted heat, as well as the gravitational effects of the Sun, planets, Moon, and over two dozen other asteroids. Bennu's spectral properties were also measured to infer its composition, forge possible connections with meteorites, and throw light on its relationship with asteroid families.

Thanks to these efforts, Bennu became the best telescopically characterized asteroid in history. Still, given its faintness, most observations could only image Bennu as a single point, like a star, rather than a resolved object like the Moon. Before the spacecraft encounter, the only spatially resolved information about the asteroid came from radar data, from which scientists created a three-dimensional model of Bennu's shape. When OSIRIS-REx finally surveyed the asteroid up close, the onboard instruments (Chapter 2) would resolve the global surface down to a scale of centimeters — an unprecedented level of detail for any planetary body.

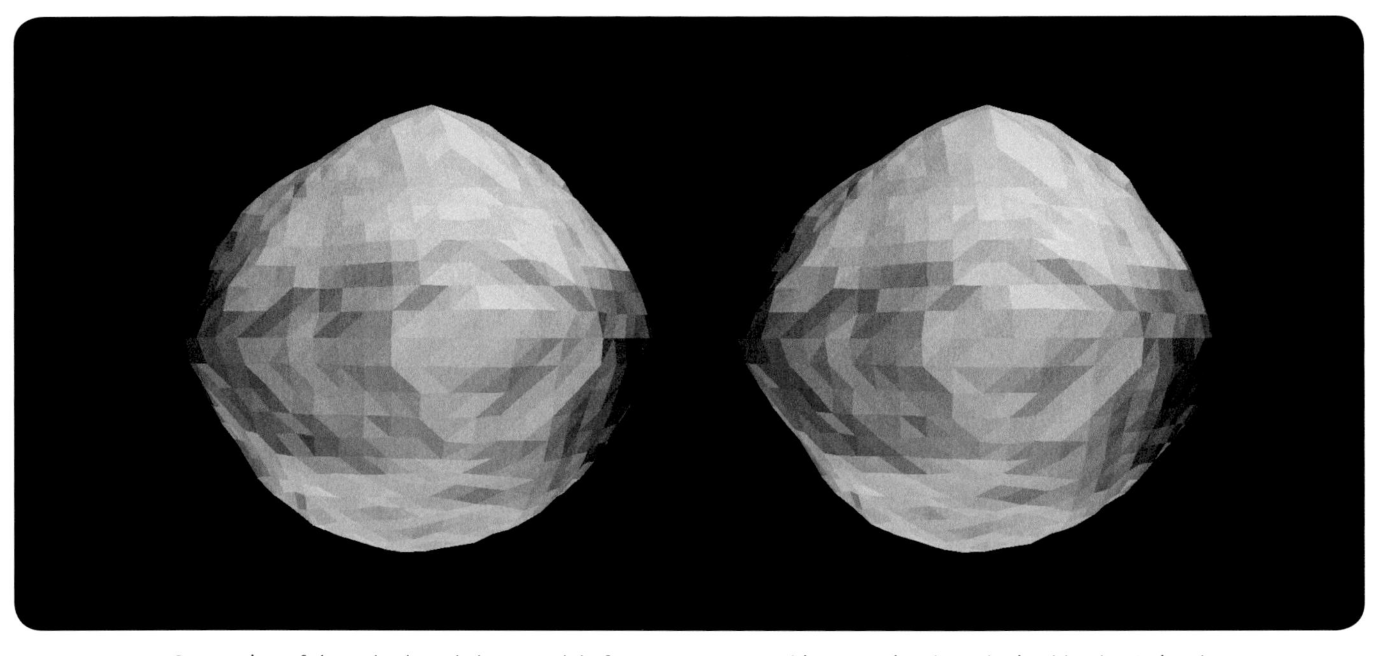

Stereo view of the radar-based shape model of Bennu, constructed in 2013 using data obtained by the National Science Foundation's Arecibo Observatory in Puerto Rico and NASA's Goldstone Deep Space Network station. Each of the triangular facets (or plates) used to approximate the shape spans about 25 m (82 ft).

How Bennu got its name

Upon its discovery, Bennu was given the provisional name 1999 RQ36. When NASA selected the OSIRIS-REx mission in 2011 with 1999 RQ36 as its target, a more inspirational name became necessary. In 2013, an international contest was held to pick 1999 RQ36's new moniker, and "Bennu" won. Bennu, illustrated opposite, is a heron-like avian deity in ancient Egyptian mythology associated with the Sun, creation, and renewal. This contest-winning name was proposed by nine-year-old Michael Puzio from North Carolina, who perceived a resemblance between the OSIRIS-REx spacecraft, with its wing-like solar panels and extended sampling arm (Chapter 2), and a heron.

The name Bennu is related to the name of the OSIRIS-REx mission. Osiris is one of the major deities in ancient Egyptian mythology. His green skin reflects his relationship with agriculture; he was believed to have established the Nile Delta as the breadbasket of ancient Egypt. He was also the judge of souls that arrived at the underworld, and pharaohs came to be regarded as his offspring. From Osiris's heart was born Bennu, depicted as a heron that rises from the ashes, like the more widely known phoenix from ancient Greek mythology.

Osiris's dual nature as both a bringer of life and a harbinger of death parallels that of carbonaceous asteroids, which may have brought the seeds of life to the early Earth yet pose the threat of immense natural disaster if they strike our planet.

The ancient Egyptian deities Bennu and Osiris.

Composition of a carbonaceous asteroid

Mission scientists used images and spectra acquired by OSIRIS-REx, together with hands-on knowledge of carbonaceous chondrite meteorites, to infer Bennu's composition before the spacecraft delivered the sample to Earth. In particular, they predicted that Bennu's surface materials are made primarily of the following:

- water-bearing clay minerals;
- carbonate minerals, similar to the salty white deposits that form when water evaporates;
- organic compounds, including some types of molecules that are constituents of life on Earth;
- and magnetite, a magnetic iron mineral.

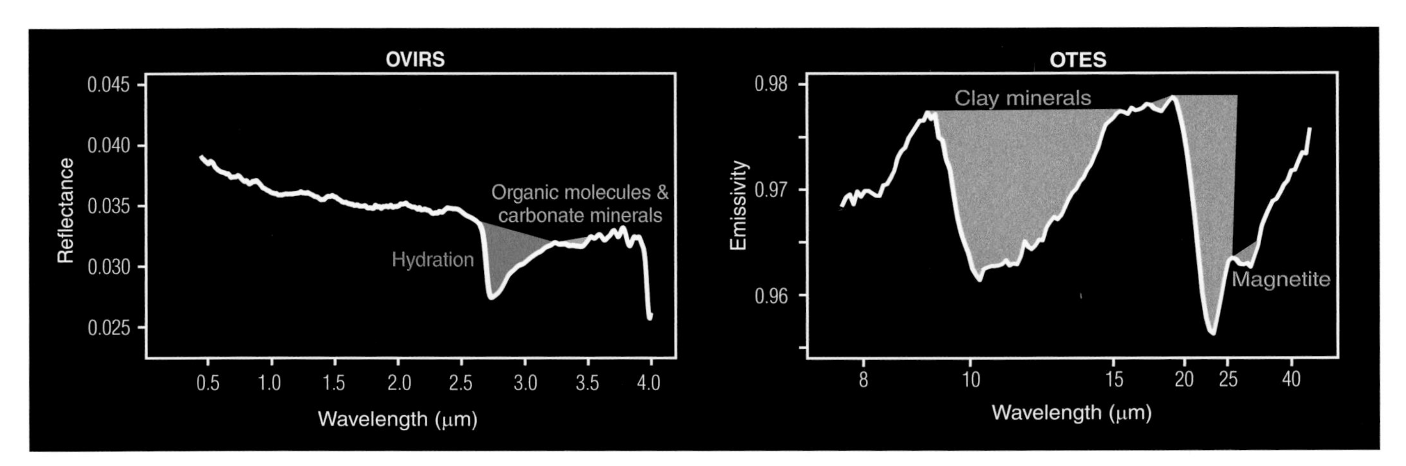

The reflected light spectrum of Bennu measured by OVIRS (left) and the thermal emission spectrum of Bennu measured by OTES (right) are shown here with annotations of the spectral peaks and valleys that indicate the asteroid's composition.

These substances point to a prior history when the component parts of Bennu's rubble were part of a much larger parent asteroid, approximately 100 kilometers (60 miles) in diameter. Bennu's minerals and organic molecules formed when liquid water percolated through the pore spaces of the original rocks, hydrothermally altering the existing phases, and precipitating new ones from solution. This composition makes Bennu an attractive target for future space mining endeavors because the water bound up in its minerals could be converted to rocket fuel. On the other hand, a very small fraction of Bennu's surface is predicted to consist of anhydrous (water-free) minerals, such as pyroxene, which may be relics of past collisions between the parent asteroid and foreign objects. The comprehensive laboratory analyses planned for the returned sample will test all these predictions.

Bennu spectrally resembles the hydrated carbonaceous chondrites, but it may not correspond wholly to any known category of meteorite in collections on Earth. This is not surprising given that the meteorite record is biased by the destructive effects of atmospheric entry. Indeed, spacecraft data suggest that the surface rocks of this asteroid are so porous and weak that a fragment of Bennu would likely disintegrate in the atmosphere before reaching the surface. This finding underscores the importance of sample return missions for understanding our neighbors in the Solar System.

Origin story

Bennu's parent probably formed beyond Jupiter in the outer protoplanetary disk, early in Solar System history. Its initial materials — including rocky, metallic, organic, and icy compounds — were inherited from the molecular cloud that preceded our Solar System, or were formed and altered in the protoplanetary disk, billions of years ago.

This large asteroid was likely heated from within by the decay of radioactive elements, until their radioactivity was exhausted about 10 million years after formation. Meanwhile, its surface was bombarded by other asteroids and space debris in the chaotic environment of the young Solar System, resulting in mixing of materials, high-pressure shock, and additional heating. All of this activity caused the ices to melt into liquid water, which reacted with the other constituents as it coursed through them, leading to the water-saturated mineralogy seen in the boulders on Bennu today.

Gravitational interaction with Jupiter evicted Bennu's parent body from the outer Solar System and launched it into the main asteroid belt. During its residence in this asteroid-dense region, it probably collided with other objects and accreted material from them, resulting in the intermixing of different asteroid compositions. Testifying to this process are very bright fragments of rock on Bennu's present-day surface that spectrally match the V-type asteroids (Chapter 7).

About one billion years ago, a catastrophic collision shattered the parent body to pieces, forming a family of smaller asteroids, including Bennu. Bennu is not a coherent monolith, like larger asteroids and planets. Rather, it is made up of loose fragments from the parent that have coalesced into a rubble pile held together by its own microgravity — approximately equivalent, at the equator, to what astronauts experience on the International Space Station.

The Yarkovsky effect gradually pushed Bennu far enough toward the Sun to reach a dynamical resonance, which then flung it out of the main belt and onto its present-day Earth-crossing orbit roughly two million years ago.

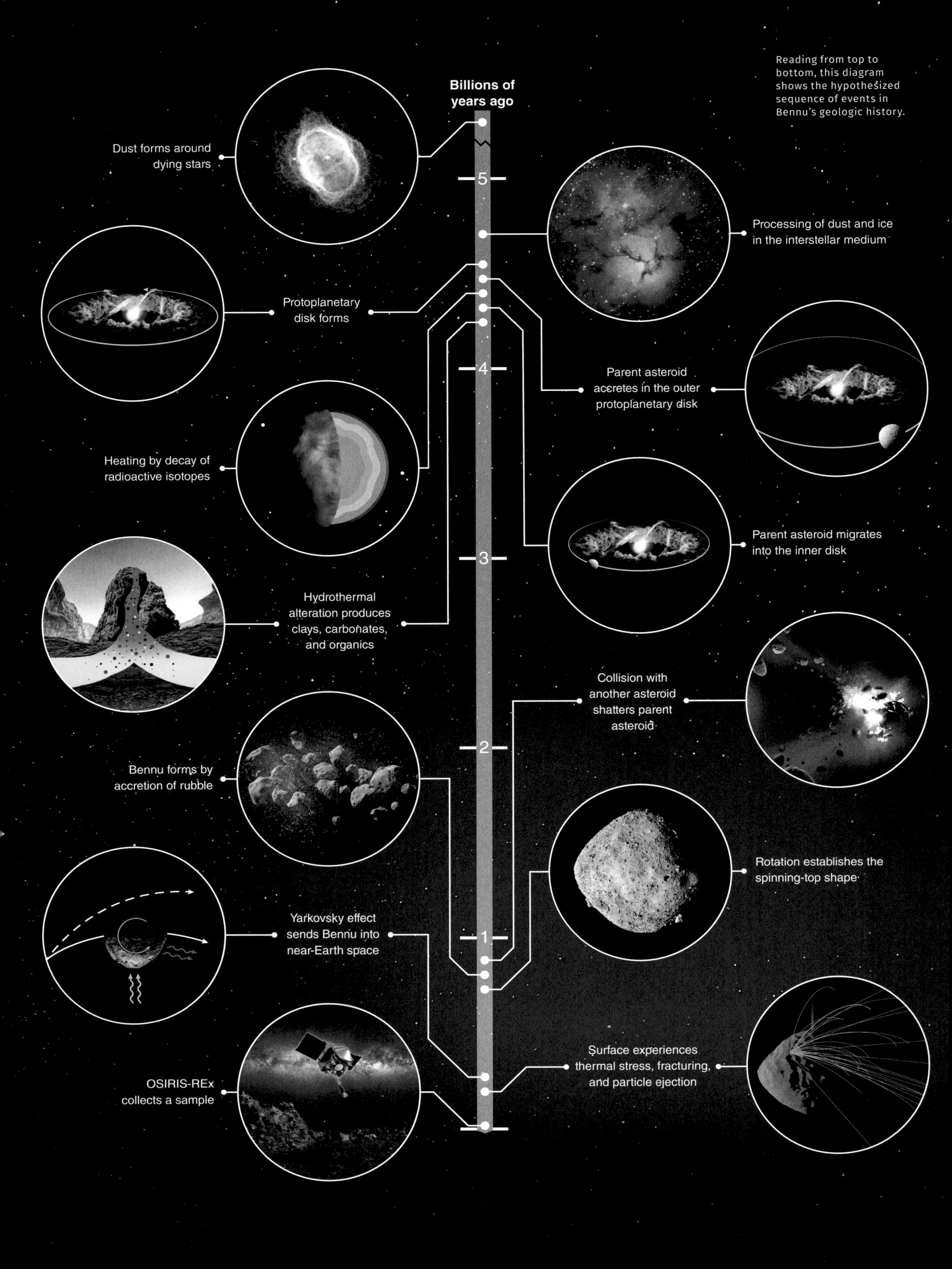

Reading from top to bottom, this diagram shows the hypothesized sequence of events in Bennu's geologic history.

Present-day position

As an Apollo-type asteroid, Bennu's orbit crosses that of Earth. Its orbit is inclined by six degrees, a small angle that required the OSIRIS-REx spacecraft to receive an energy boost from Earth's gravity en route to the asteroid.

Each Bennu year, or complete revolution around the Sun, is equivalent to 1.2 Earth years. However, Bennu's orbital path is evolving; it shrinks by about 285 meters (935 feet) in semi-major axis per year as a result of the Yarkovsky effect.

In the vastness of space, 285 meters is a trivial distance. But these slight changes in the orbit accumulate over time, shortening Bennu's year and, most importantly, creating uncertainty about whether Bennu will impact the Earth in the future.

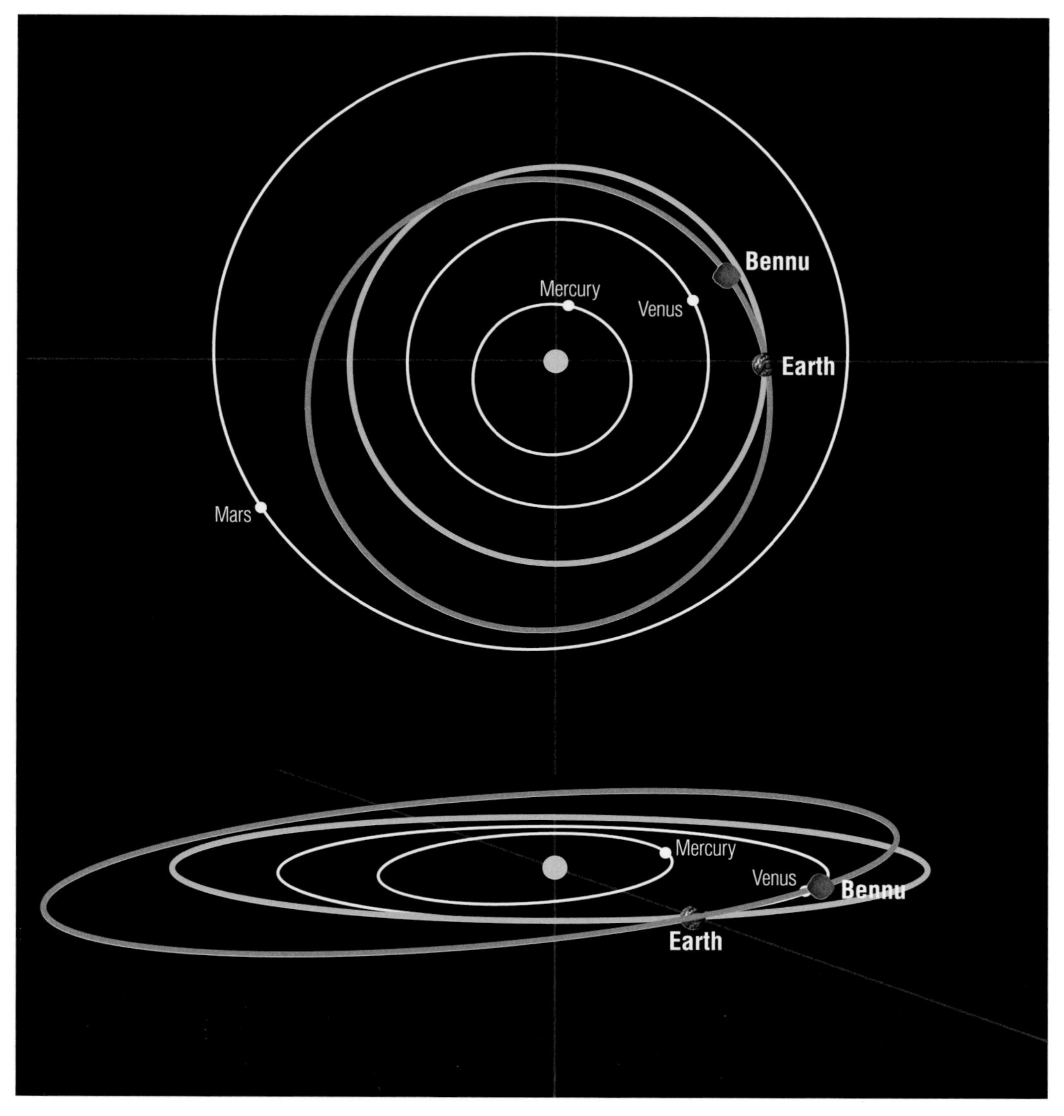

Bennu's orbital path and position on the date of sample return (September 24, 2023). The top diagram shows Bennu's orbit viewed from celestial north. The bottom diagram illustrates its ~6° inclination from the ecliptic plane.

Future trajectory and potential Earth collision

Bennu's Earth-crossing orbit poses the risk of a collision with our planet in the 22nd century. Thanks to the OSIRIS-REx encounter, the Yarkovsky effect has been measured to very fine precision, but multiple outcomes of Bennu's gradual inward drift are still plausible. The orbit is well understood until the year 2135. But, in September of that year, Bennu will pass between the Earth and the Moon. Due to intricacies of the gravitational interactions during this encounter, there are large uncertainties in our knowledge of Bennu's subsequent trajectory. The specific details of that encounter will determine Bennu's fate. In most scenarios, Bennu misses our planet and continues its journey around the Sun. But in other outcomes, Bennu enters a trajectory to strike Earth, with the most likely impact occurring on September 24, 2182. The cumulative probability of any Earth impact through the year 2300 is small — 1 in 1,750 — but it is large enough to put Bennu in first place on NASA's Sentry index of potentially hazardous space objects. From the perspective of life on Earth, Bennu is the Solar System's most dangerous asteroid.

If Bennu winds up on a collision course with our planet, it will slam into Earth at a velocity of 12 kilometers (7.5 miles) per second, releasing three times more energy than all nuclear weapon detonations in history. The enormity of such an event is difficult to grasp in human terms, but Earth has weathered roughly 40,000 impacts of similar magnitude over the past 4 billion years. This collision would leave a scar, in the form of a crater several kilometers wide and 500 meters (547 yards) deep.

Nevertheless, it is much more probable that Bennu will bypass Earth. As its orbit contracts, there is a small chance of a later collision with Venus or Mercury. The most likely outcome is that Bennu will continue to spiral inward until it falls at last into the Sun.

First glimpses from OSIRIS-REx

One of the most exciting phases of any space mission is seeing the destination emerge from the blackness of space. The numerous observations previously made from Earth are verified or refined by the details that gradually unfold, while other discoveries are unexpected.

As the OSIRIS-REx spacecraft made its approach to Bennu in late 2018, it gave us a view of the asteroid evolving from a pixelated blur to a resolved world, filled with rugged, compelling landscapes. For the team, the thrill of this gradual revelation was alloyed with apprehension. The mission was designed to collect a sample from an open sandy area, which pre-launch telescopic data had indicated should be readily available on Bennu. The images returned by the spacecraft, however, revealed a rough surface littered with boulders, many the size of buildings, with no sandy touchdown options in sight. The unexpectedly rocky landscape presented the team with a difficult road to sample acquisition (Chapter 5).

This surprise also opened up a galvanizing new line of scientific inquiry — why is Bennu so different from what we expected, and what are the implications for what we know, or think we know, about other asteroids?

Stereo view of Bennu as seen by OSIRIS-REx's PolyCam imager on October 27, 2018, from a range of about 3,000 km (1,850 mi). Some possible surface features and/or variations in albedo (brightness) are faintly visible. Bennu's north pole is up in this and subsequent images, unless otherwise specified.

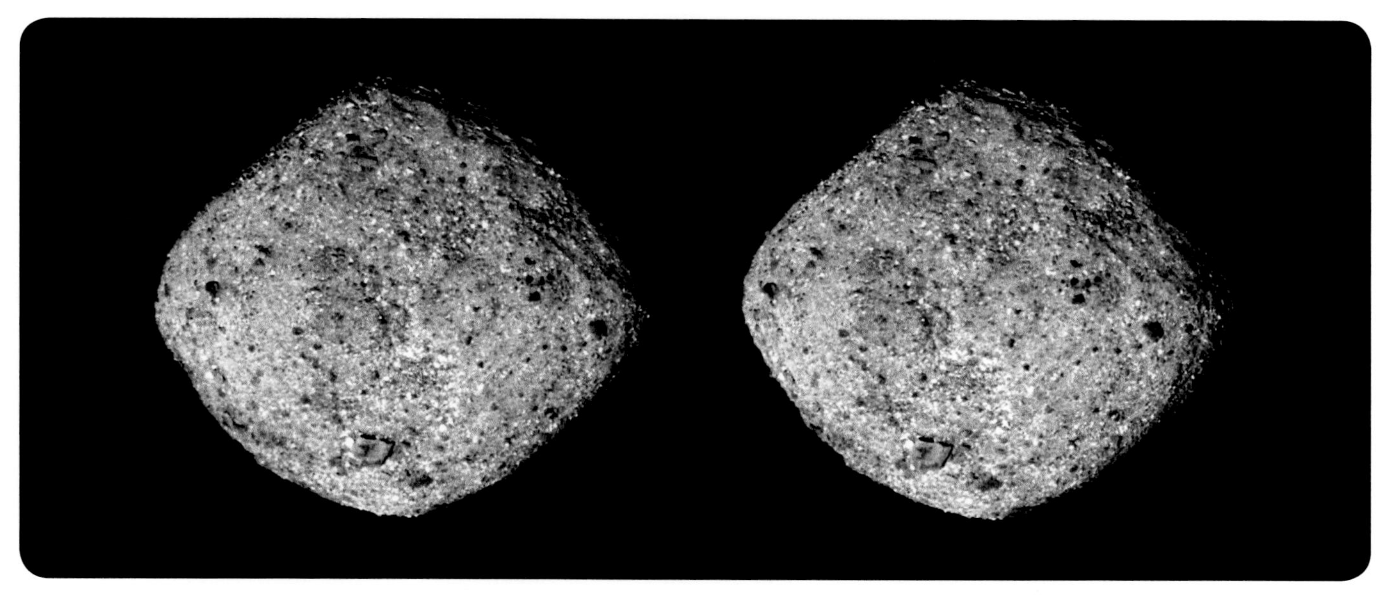

Stereo view of fully illuminated Bennu, acquired on November 25, 2018, at a range closer to 100 km (60 mi). This view was risky to obtain because it required the Sun to be directly behind the spacecraft, which might have put critical components at risk of overheating. It captures Bennu at its least shadowed, highlighting the variety of albedos on the surface, as well as many large boulders.

The view sharpened further when the spacecraft rendezvoused with Bennu and flew over the poles and equator. Acquiring images with different viewing angles and lighting conditions brought the rims of large craters into stark relief, and long shadows indicated that some boulders loom tall over the surface.

Although the surface of Bennu turned out to be very different from expectations, its physical dimensions, as measured by the spacecraft, agreed well with predictions from the ground. In particular, Bennu has a diamond, or spinning-top, shape. Its equatorial diameter is much larger than the distance from pole to pole — consistent with the prediction of the coarse radar-based model shown earlier in this chapter.

We now know that this characteristic shape is shared with other rubble piles, such as Ryugu (Chapter 2). This shape might have formed right away when the rubble pile accumulated after the catastrophic disruption of its parent body. Alternatively, it may be the result of gradual changes in how fast an asteroid spins on its axis. Bennu is already rotating rapidly; one Bennu day, or complete rotation, is just 4.3 hours long. In addition, the YORP effect (Chapter 1) is causing it to spin more and more rapidly. Over hundreds of thousands of years, YORP-driven acceleration of the rotation can cause rubble to redistribute via surface creep or landslides. This redistribution alters the surface slopes and thereby the overall shape of the asteroid. Although the equator bulges outward, centrifugal forces make it the geopotential low. The result is, counterintuitively, that loose surface material tends to flow uphill towards it. Thus, past periods of acceleration may have produced the equatorial bulge as material migrated from the poles and midlatitudes toward the equator (Chapter 7).

Material at Bennu's equator is moving at close to orbital speeds. If a person released a rock above their head at the equator, it would become a natural satellite. Bennu is thus very close to terminal instability. As its rotation rate increases, it may fling off a disk of material and restabilize, or completely disintegrate.

This mosaic of Bennu was created from 12 PolyCam images taken on December 2, 2018, when OSIRIS-REx approached to around 24 km (15 mi) from the surface. BenBen protrudes from the lower right.

Acquired by MapCam on December 11–12, 2018, as the spacecraft closed in for an equatorial flyover. BenBen can be seen in the southern hemisphere, close to the south pole at about 7 km (4 mi) range.

Acquired on December 13, 2018, during the equatorial flyover. This view shows much of the same surface as the preceding stereo pair, rotated slightly left and illuminated from the opposite side.

A different view as the spacecraft departed from the equatorial flyover.

Acquired on December 9, 2018, looking down at the north pole. Both poles remain shadowed at all times because Bennu's axis of rotation is nearly perpendicular to the Sun, unlike Earth, whose 23° axial tilt illuminates the poles in opposite seasons.

Acquired on December 16, 2018, looking down at the south pole.

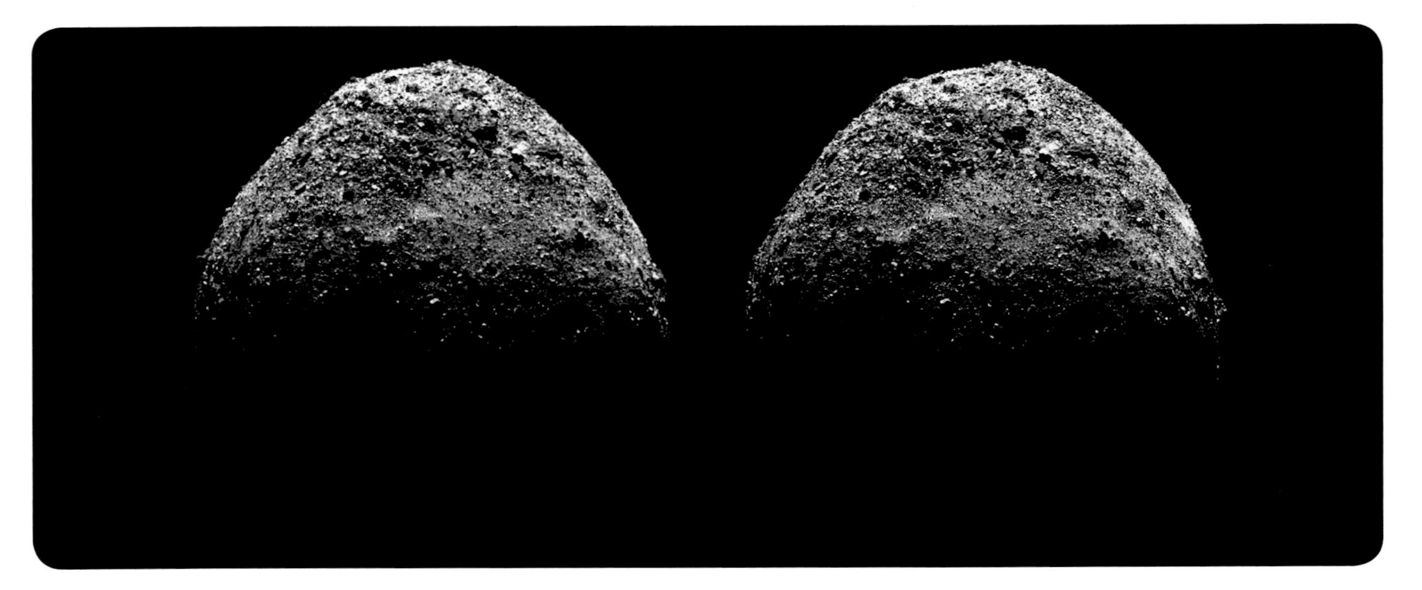

Acquired on December 17, 2018, while receding from the south pole flyover.

Particle ejection events

Part of any space mission is checking for hazards in the vicinity of the destination. The searches for dust plumes and moonlets as the spacecraft approached had shown nothing. But navigation images collected after the spacecraft entered its first orbit in January 2019 had a surprise in store. The team noticed star-like points of light where no stars should be. These were soon discovered to be pebble-sized, flaky fragments of rock ejected from various locations on the asteroid's surface.

The mission's scientists and engineers worked together to understand the behavior of these projectiles. Fortunately, the particles were found to be too small and too slow-moving to pose a significant risk to the spacecraft or its instruments. The team applied techniques normally used in navigating the spacecraft to reconstruct the flight paths of the particles, based on their changing positions in a series of images.

Of the numerous ejected particles that OSIRIS-REx observed over the year that followed, about two-thirds fell back to Bennu's surface after being launched, and the remaining third either escaped into space or orbited Bennu for up to a few days. One particle was observed to ricochet off Bennu's surface and re-launch. Ejection events predominantly occurred in the afternoon and evening (local time) and from low latitudes but seemed to be able to occur at any time and from anywhere. The series of stereos overleaf was derived from reconstructions of the particle trajectories.

What could cause this phenomenon? The most likely culprits are related to Bennu's lack of an atmosphere. Without this protective shell of air, Bennu is subjected to extreme day/night temperature variations, sometimes exceeding 100 degrees Celsius at the equator. The dramatic temperature shifts put a mechanical strain on the rocks, eventually causing them to crack. The fracturing could release enough energy to propel splinters of rock into space. In addition, very small pieces of rocky space debris, or meteoroids, constantly bombard Bennu's unprotected surface at energies sufficient to launch loose particles. A third possible mechanism is dehydration of water-rich clay minerals, causing expansion along their sheet-like crystal planes that eventually blasts them apart. These processes could be working separately or together to activate Bennu's surface.

Although Bennu's rugged terrain posed severe operational challenges to the OSIRIS-REx team, its mysteries make it an asteroid geologist's dreamscape. The next chapter is a tour of the most fascinating areas on Bennu's surface that helped scientists unravel its history.

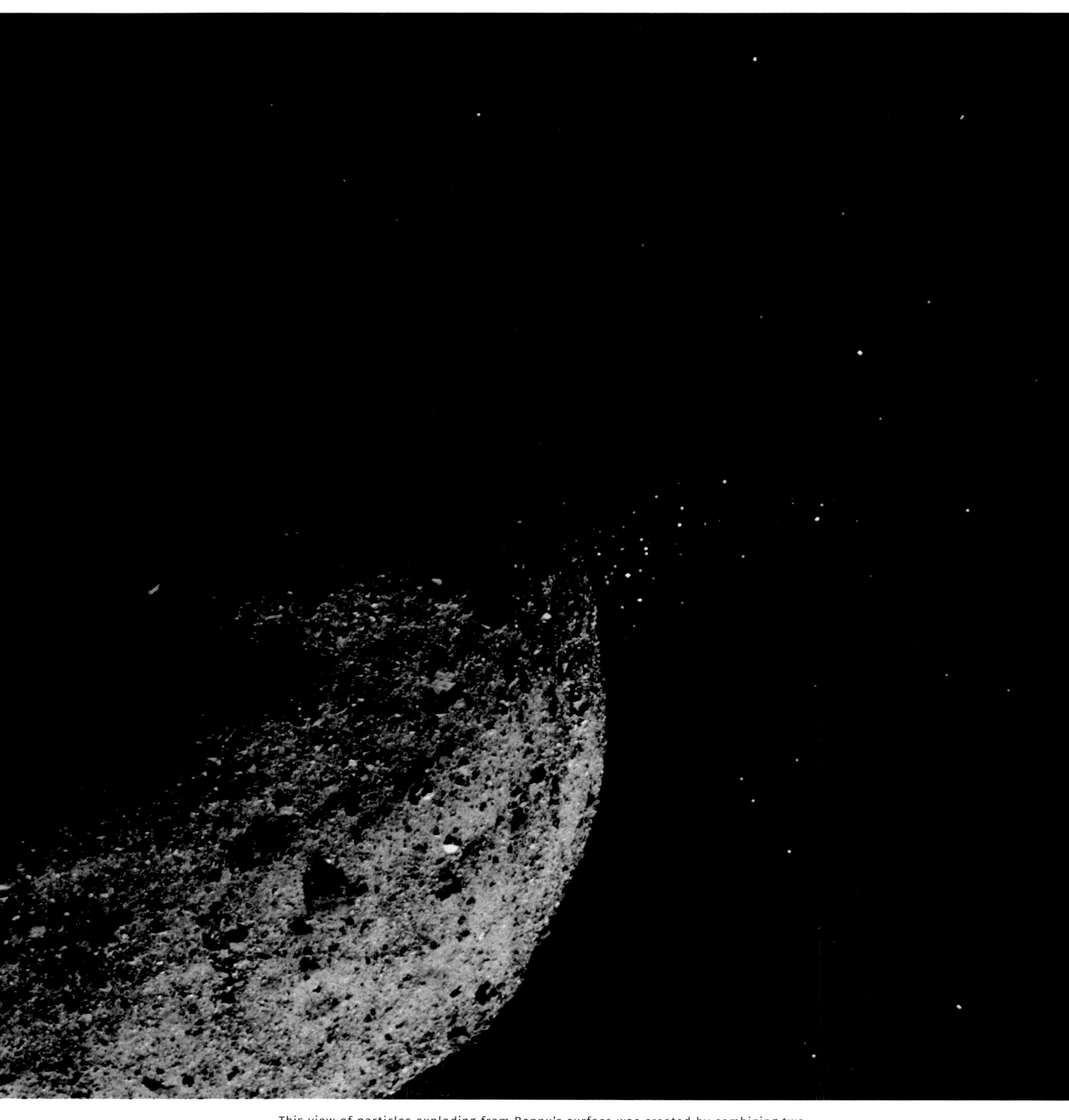

This view of particles exploding from Bennu's surface was created by combining two images taken by NavCam 1 on January 19, 2019: a short-exposure image (1.4 ms), which shows the asteroid, and a long-exposure image (5 s), which shows the particles.

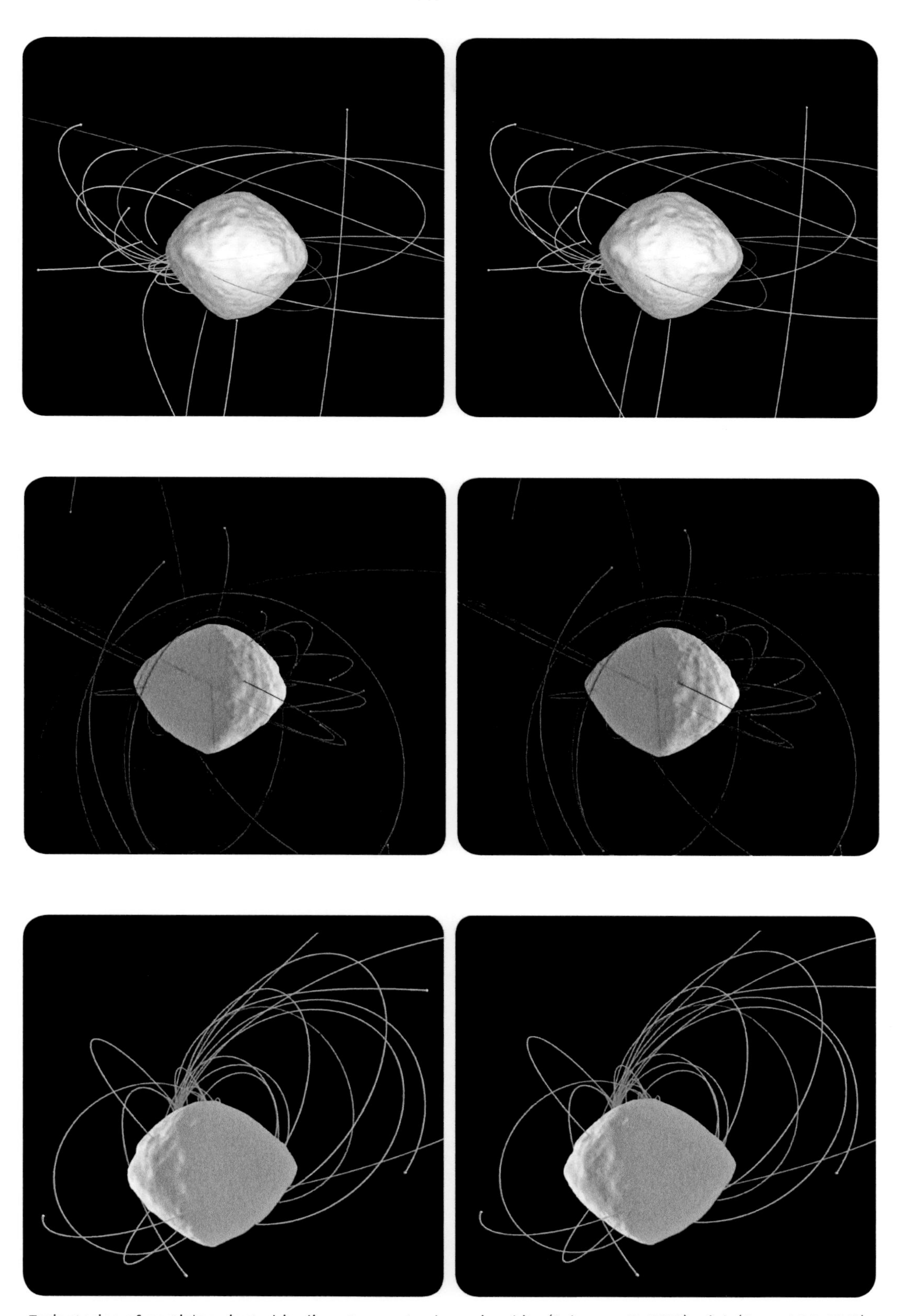

Trajectories of particles ejected in discrete events shown in white (February 11, 2019), pink (August 28, 2019), and green (September 14, 2019). This is an illustration of real data, not an artistic conceptualization. Many of the trajectories trace back to the same location on the asteroid, indicating a simultaneous launch from a single origin.

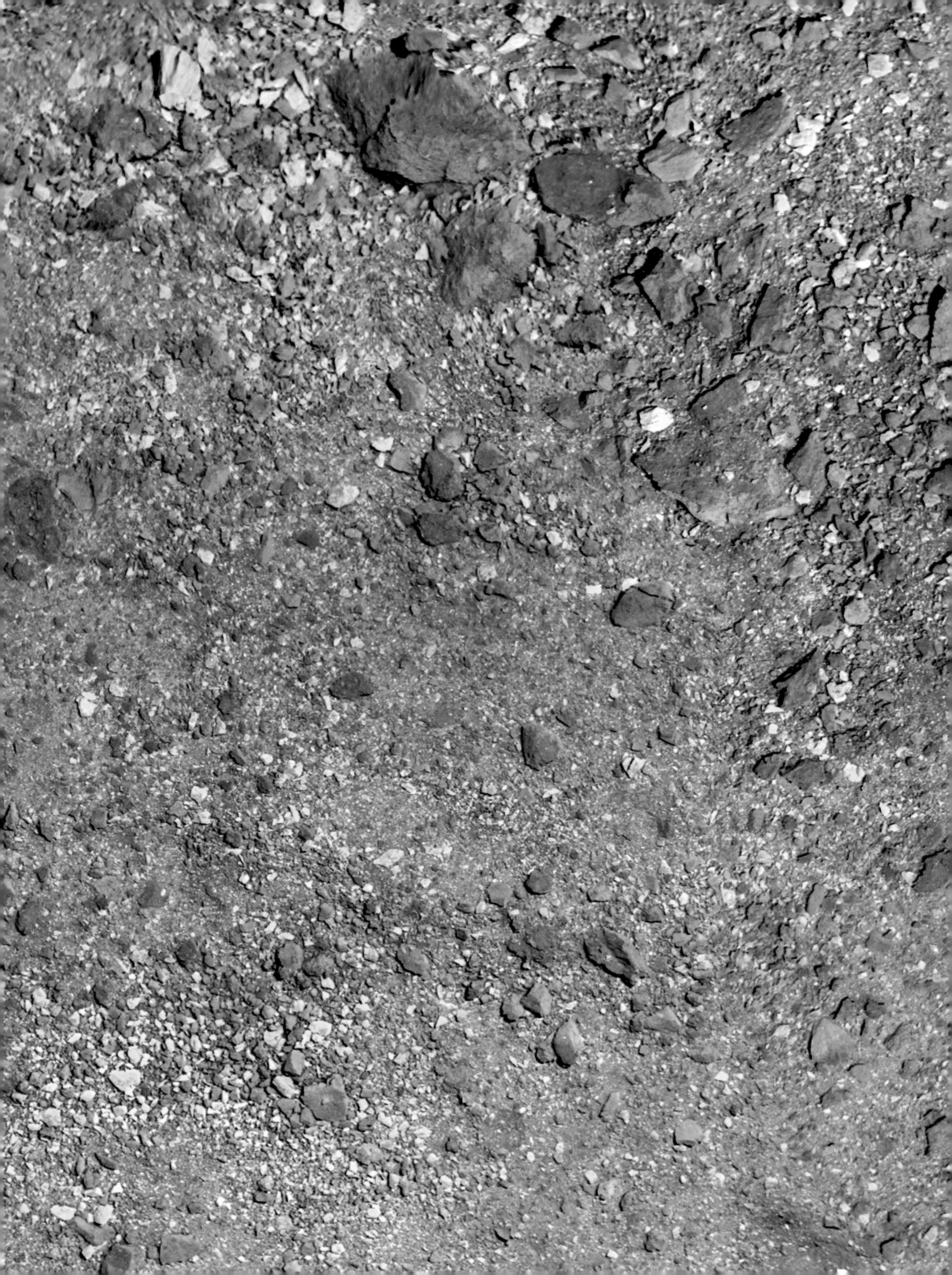

Chapter 4

Local Landmarks

Over the course of the mission, the OSIRIS-REx team became closely acquainted with the geologic features of Bennu's terrain. Large craters with pronounced rims dent the equator, while smaller craters are sometimes barely visible in images, indicated only by subtle differences in topography. Boulders come in a range of textures, shapes, colors, and sometimes daunting dimensions. In this chapter, these landmarks emerge in high resolution and three dimensions.

The OSIRIS-REx team applied to the International Astronomical Union (IAU) to officially name features of particular scientific interest and/or large size. Given that Bennu was named for a heron-like deity from Egyptian mythology, the nomenclatural theme for its geologic features is mythological birds, birdlike creatures, and places associated with them. This theme befits the mission's emphasis on the origins of life and our planet.

Here we focus on the 36 features named during the mission. Many features on Bennu's surface have yet to be named, and anyone may apply to the IAU to do so. Supporting references for the mythological origins of the names are documented in the US Geological Survey's *Gazetteer of Planetary Nomenclature*, which we follow here.

Throughout this chapter, yellow crosses in mono (2-D) images indicate the center of the feature of interest. The features are presented approximately in order of increasing longitude, that is, roughly from west to east. Center coordinates and diameters of features are tabulated in the Features Index.

The 3-D orientations of stereo images are indicated by compass icons in the margins, wherein the north arrow is larger when pointing toward the reader and smaller when pointing away.

(*Opposite*) This close-up in false color of a 280 m (920 ft) wide area in Bennu's southern hemisphere highlights the diversity between — and even within — its boulders. Areas colored redder have more positive spectral slopes in the MapCam wavelengths, and those colored bluer have more negative spectral slopes (see Chapter 2 and the spectral color map in Chapter 7 for details). The variation in color results from the boulders' different starting compositions and exposure histories. An ovoid rock in the bottom center has a strikingly blue, recessed face. This may have been recently cleaved by thermal fracturing and its surface less space-weathered than the adjacent, lighter blue face. Another rock in the center left grades from red to blue, probably because the bluer, equator-facing side is more exposed to weathering. This rock, called Amihan Saxum, is situated in Tlanuwa Regio, and both appear later in this chapter. For location, see the map of named features on the next page.

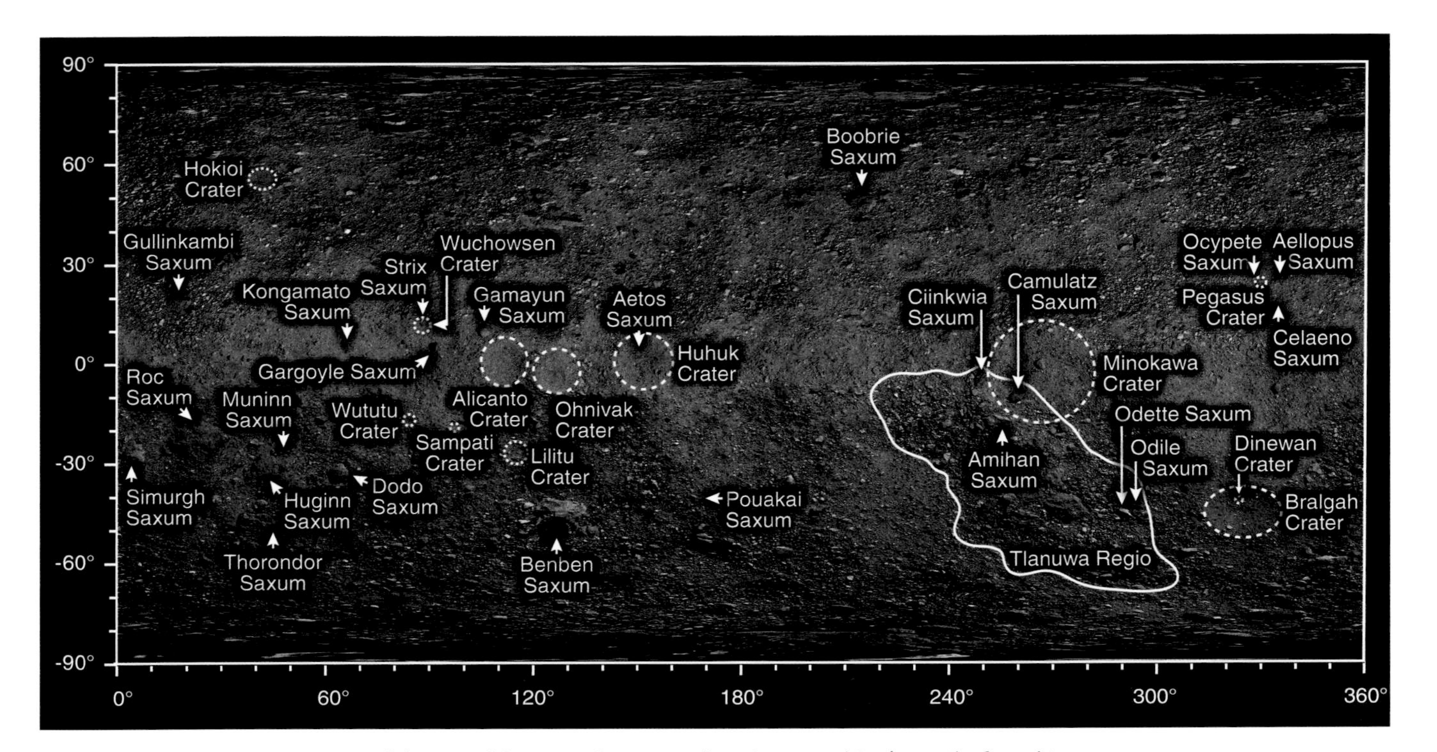

Map of the named features of Bennu, to date. Saxum and Regio are the formal terms for "boulder" and "broad geographic region", respectively.

Some of Bennu's largest boulders march across this view of the asteroid's southwestern quadrant. From approximately west to east: Simurgh, Roc, Huginn, Muninn, Dodo, and, protruding well above the horizon, BenBen. Each of these boulders is shown in more detail in this chapter.

Simurgh Saxum

Named after a benevolent bird from Persian mythology who possesses all knowledge, Simurgh Saxum defines the prime meridian on Bennu and is the basis for the asteroid's coordinate system. It was chosen for this purpose owing to its large size of almost 40 m (130 ft) and distinctive angular shape, which make it easy to spot from nearly any angle. In the mono image, Simurgh is viewed from the north, looking south over the equator.

N

Field of view,
160 m

Roc Saxum

With an exposed length of about 100 m (328 ft) (comparable to a rugby pitch or American football field), and possibly an even larger extent below the surface, Roc Saxum is the longest boulder on Bennu and appropriately named after an enormous bird of prey in Middle Eastern mythology. Roc Saxum showed up in early OSIRIS-REx images as a dark patch so huge that it was not initially identified as a single coherent object.

Roc epitomizes the primary scientific paradox of Bennu: Prior to OSIRIS-REx's arrival, large boulders on asteroids were thought to heat up and cool off very slowly (high thermal inertia). However, Roc Saxum heats up and cools off more quickly (low thermal inertia) than any other location on the surface, the opposite of expectations (see Chapter 7 for more on Bennu's thermal properties).

In the stereo image, Roc is viewed from above, and Simurgh can be seen to its left. The mono image views Roc obliquely from the north looking south.

Field of view,
200 m

Gullinkambi Saxum

Named after the gold-combed rooster from Norse mythology who lives in the heavenly hall Valhalla and wakes gods and heroes, Gullinkambi Saxum towers above the surface of Bennu at a height of 20 m (65 ft). Its distinctive layers hint at flow and deposition of sediment on Bennu's parent asteroid. In the mono image, coarser clasts embedded in the layers can be faintly seen.

Field of view,
35 m

Hokioi Crater

Hokioi Crater is among the spectrally redder-than-average craters on Bennu, suggesting comparatively unweathered material and thus a young age. This 20 m (65 ft) diameter crater (about the length of a tennis court) contains the site dubbed Nightingale from which the OSIRIS-REx spacecraft eventually collected its sample (Chapters 5 and 6). Hokioi is a Māori (New Zealand) mythical bird with a long divided tail who lives in the heavens.

N

Field of view, 35 m

Huginn and Muninn Saxa

Huginn and Muninn Saxa, about 50 and 30 m (164 and 98 ft) long, respectively, are named for two ravens in Norse mythology that accompany the supreme god Odin. On Bennu these two saxa lie next to each other and may be two different surface expressions of a larger, partially buried boulder.

On Huginn Saxum, a slab can be seen protruding upward in the stereo image. This partially separated slab suggests exfoliation fracturing, where flakes of a boulder break off owing to thermal stresses from Bennu's rapid day-night temperature cycling.

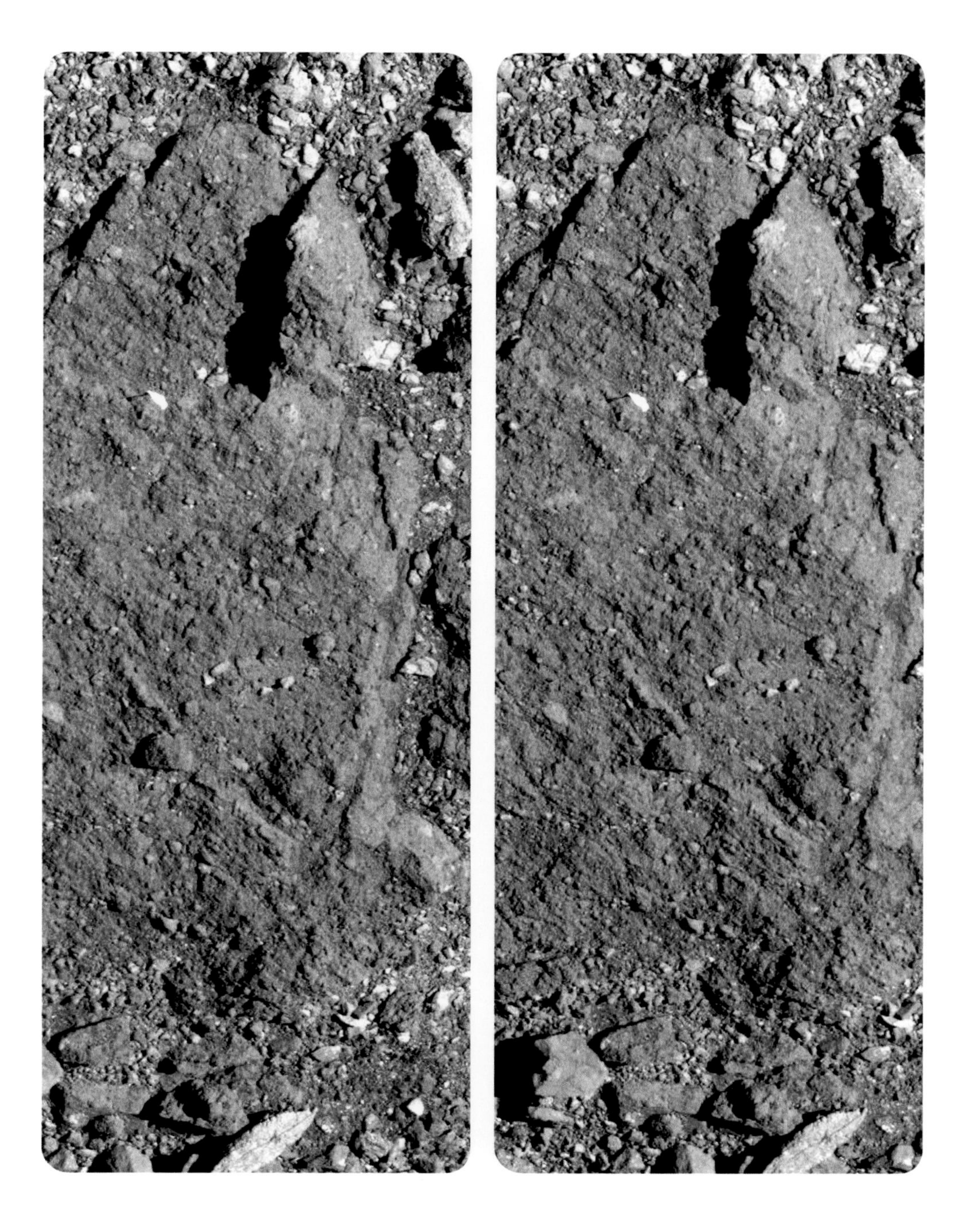

N

Field of view,
20 m

Thorondor Saxum

This dark boulder is almost completely buried but has an apparent “wingspan” of about 55 m (180 ft) — the same as its namesake, Thorondor, the King of the Eagles in fantasy novels by English author J.R.R. Tolkien. The rock fragments across its surface are likely migrating northward (toward the upper left corner in the mono image) in the direction of the equator, owing to the centrifugal forces from Bennu’s rotation.

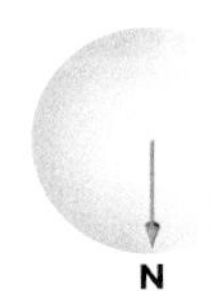

Field of view, 25 m

Dodo Saxum

Because of its circular shape, Dodo Saxum was named after the extinct, flightless and particularly rotund dodo bird, who is also a fictional character in the book *Alice's Adventures in Wonderland* (1865) by English writer Lewis Carroll. Dodo Saxum exhibits a change in color and texture across a linear boundary, suggesting different formation environments on Bennu's parent asteroid.

N

Field of view, 30 m

Kongamato Saxum

Kongamato, or "breaker of boats," is a pterosaur-like creature in Kaonde myths (northwest Zambia and adjacent areas). This dark boulder is strewn with fragments of brighter material — like pieces of a shattered boat — which provide evidence that some of the particles launched off the surface by impacts or thermal fracturing (Chapter 3) fall back onto Bennu.

N

Field of view,
25 m

Wututu Crater

This is an example of a small, difficult-to-identify crater. Like many other such craters on Bennu, Wututu Crater lacks obvious signifiers such as a sharp rim or ejecta deposit. However, its interior appears to have smaller surface particles than its surroundings, hinting at an impact-derived origin. The subtlety of this feature made its designation as a crater a subject of debate among OSIRIS-REx scientists. The stereoscopic image below clearly demonstrates that it is a topographic depression. Wututu Crater was named for the Fon (Benin) mythical bird who reconciled a dispute between two gods and thereby ended a drought.

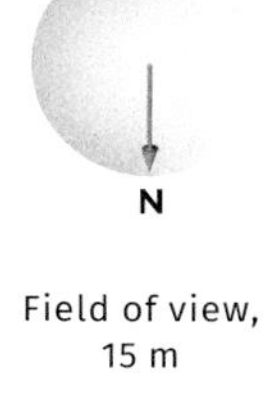

Field of view, 15 m

Strix Saxum in Wuchowsen Crater

Wuchowsen Crater, 20 m (65 ft) diameter, is the best example on Bennu of a classic bowl-shaped crater and was to be the location of the mission's backup sample site, nicknamed Osprey (Chapter 5). Wuchowsen is a giant bird in Abenaki (Algonquian, northeastern North America) lore who produces winds with his wings. Strix Saxum, named after a Roman vampiric bird of ill omen, was nicknamed "twelve o'clock rock" by the OSIRIS-REx team owing to its location at the northern edge of the crater, where it resembles a tick mark on a watch face. The stereo image reveals a smooth, apparently polished surface, raising puzzling questions about weathering on Bennu or its parent asteroid.

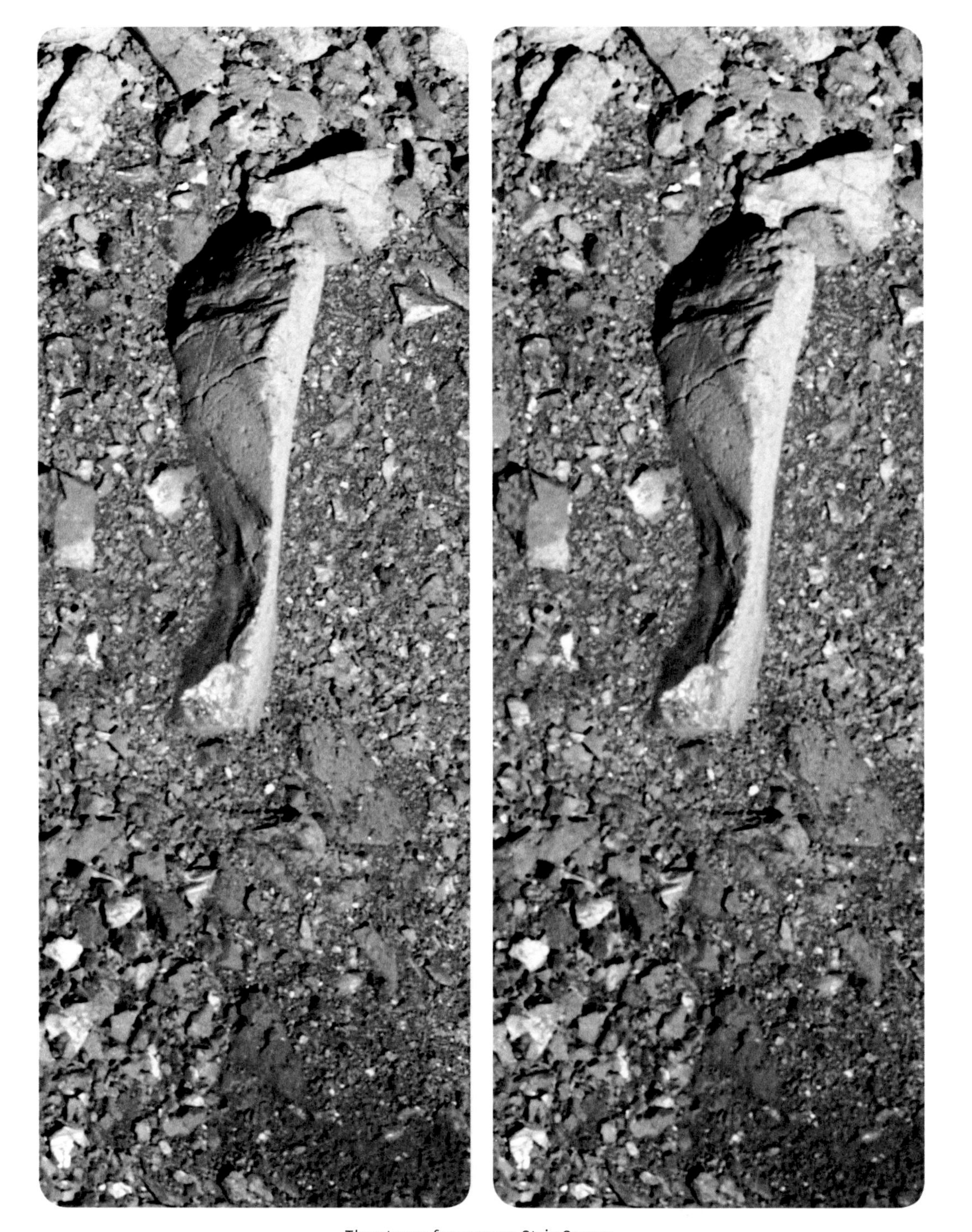

Field of view,
5 m

The stereo focuses on Strix Saxum.

Gargoyle Saxum

Gargoyle Saxum was named after the French dragon-like monster whose likeness can be found carved from stone atop old buildings. Gargoyle Saxum is one of Bennu's most striking features, 15 m (49 ft) tall, about the height of the Hollywood Sign. OSIRIS-REx scientists often used clues to understand the shapes of boulders captured in a limited number of images: for example, the long shadow Gargoyle casts, compared to boulders around it, bespeaks its lofty height. In the oblique view, two distinct surface textures can be seen: coarse and fine grained, both apparently layered. The brighter rock resting on Gargoyle is about as long as a person is tall.

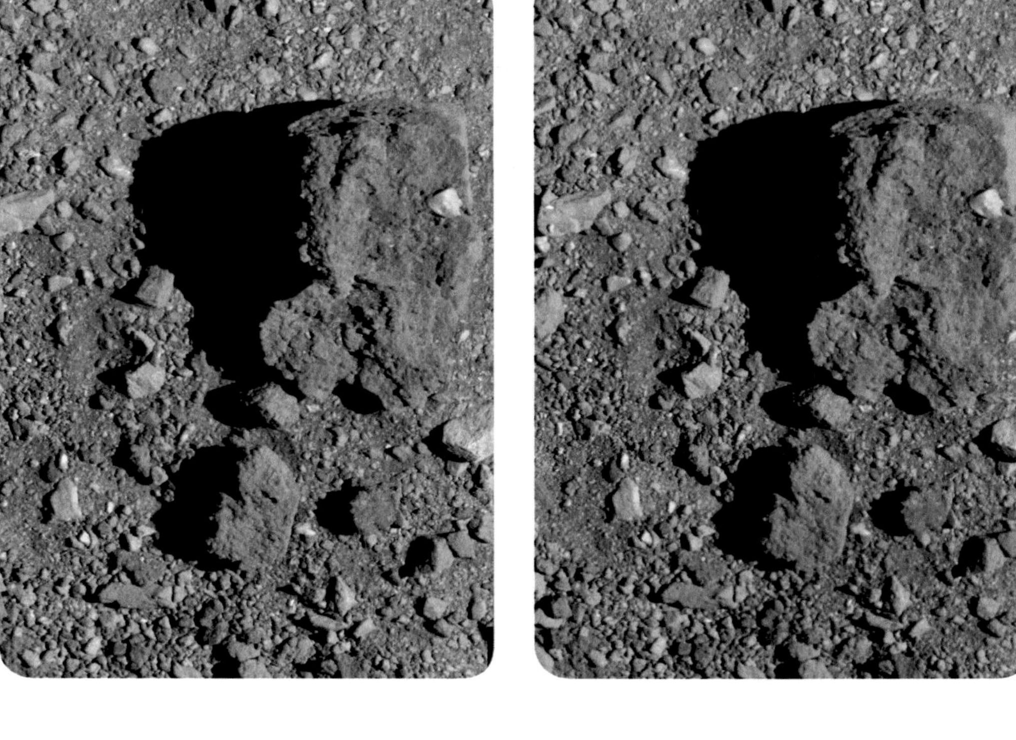

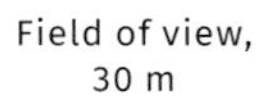

Field of view, 30 m

Sampati Crater

Similar to Wututu Crater, Sampati Crater is a small, subtle impact feature on Bennu. The stereo view below exaggerates the vertical scale due to its long baseline between left and right images. This distorts the apparent shape of the rocks, but nicely highlights the crater shape. In Hindu mythology, Sampati is a vulture whose wings were burnt when he spread them to protect his brother, who had flown too close to the Sun.

N

Field of view, 30 m

Gamayun Saxum

This boulder appears perched on the surface of Bennu. It shows extensive surface fracturing implying planes of weakness throughout. Gamayun is a prophetic bird from Slavic mythology with a woman's head, a symbol of wisdom and knowledge.

N

Field of view, 25 m

Alicanto and Ohnivak Craters

These two large craters on Bennu's equator have distinctive, pronounced rims. In a microgravity environment lacking an atmosphere, such as Bennu's, a given impact will make a larger crater in a weak material than in a strong one. The surface of Bennu is very weak, meaning that large craters like these can form as a result of relatively small impacts. The oldest and largest craters on Bennu could be less than 100 million years old — young in geologic terms.

Alicanto is a bird from Chilean mythology whose wings shine at night with metallic colors derived from the metal ores it eats. Ohnivak is a glowing firebird from Czech and Slovak folklore with red, gold, and orange feathers.

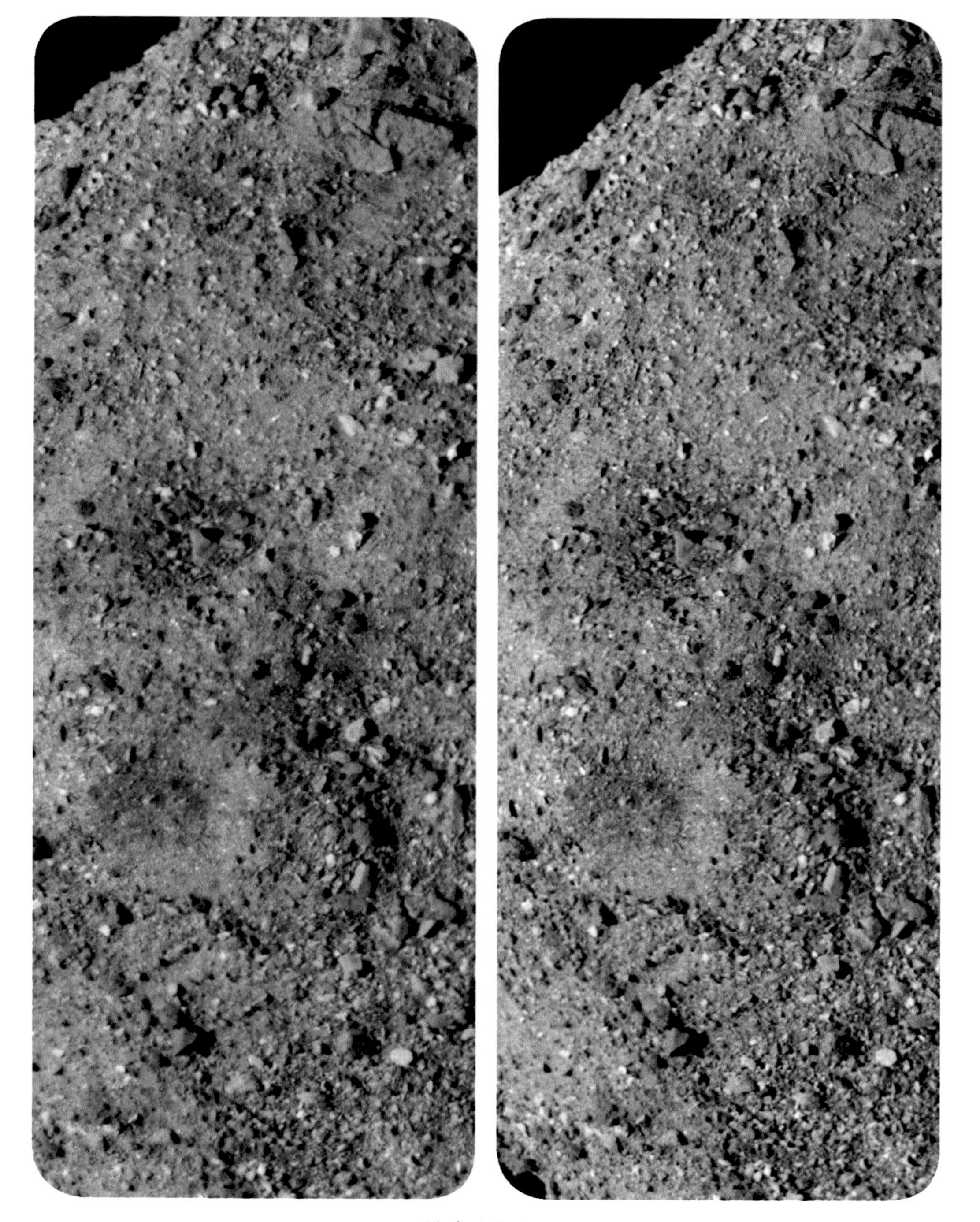

Ohnivak Crater.

N

Field of view,
300 m

Alicanto (center) and Ohnivak (right).

Alicanto (top) and Ohnivak (below). North is to the right.

Lilitu Crater

This crater has several smaller craters on and near its rim. The smaller craters can be considered younger than Lilitu Crater because they overprint it. Lilitu is a Mesopotamian/Sumerian nocturnal demon depicted as a winged woman with the feet and talons of a bird.

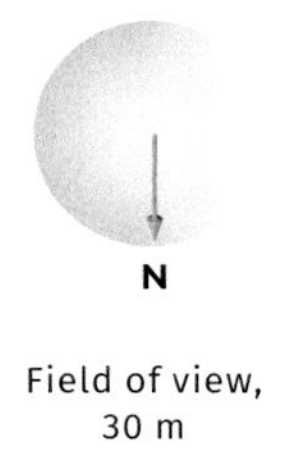

Field of view, 30 m

Aetos Saxum

This boulder, named for a monstrous eagle from Greek mythology, has a distinctive flat and platy shape, suggesting high strength compared to the average boulder on Bennu. Aetos Saxum is located in the lower half of the field of view in the stereo image. It is slightly less than 15 m (50 ft) long.

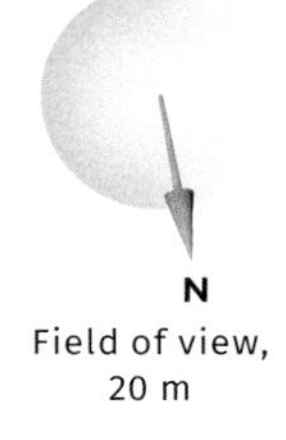

Field of view, 20 m

BenBen Saxum

BenBen Saxum is the tallest boulder on Bennu at nearly 40 m (130 ft) in height. It was the only identifiable feature from the ground-based telescopic data and forms an iconic bump in Bennu's profile. In the mosaicked image at the bottom of the page, the field of view is about 390 m (1,280 ft).

BenBen Saxum is named after the ancient Egyptian mythological mound that arose from the primordial waters. The god Atum, an antecedent of the god Osiris, settled on it to create the world after his flight over the waters in the form of the Bennu bird.

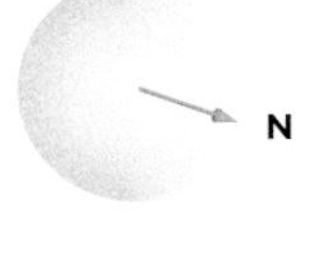

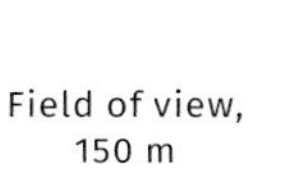

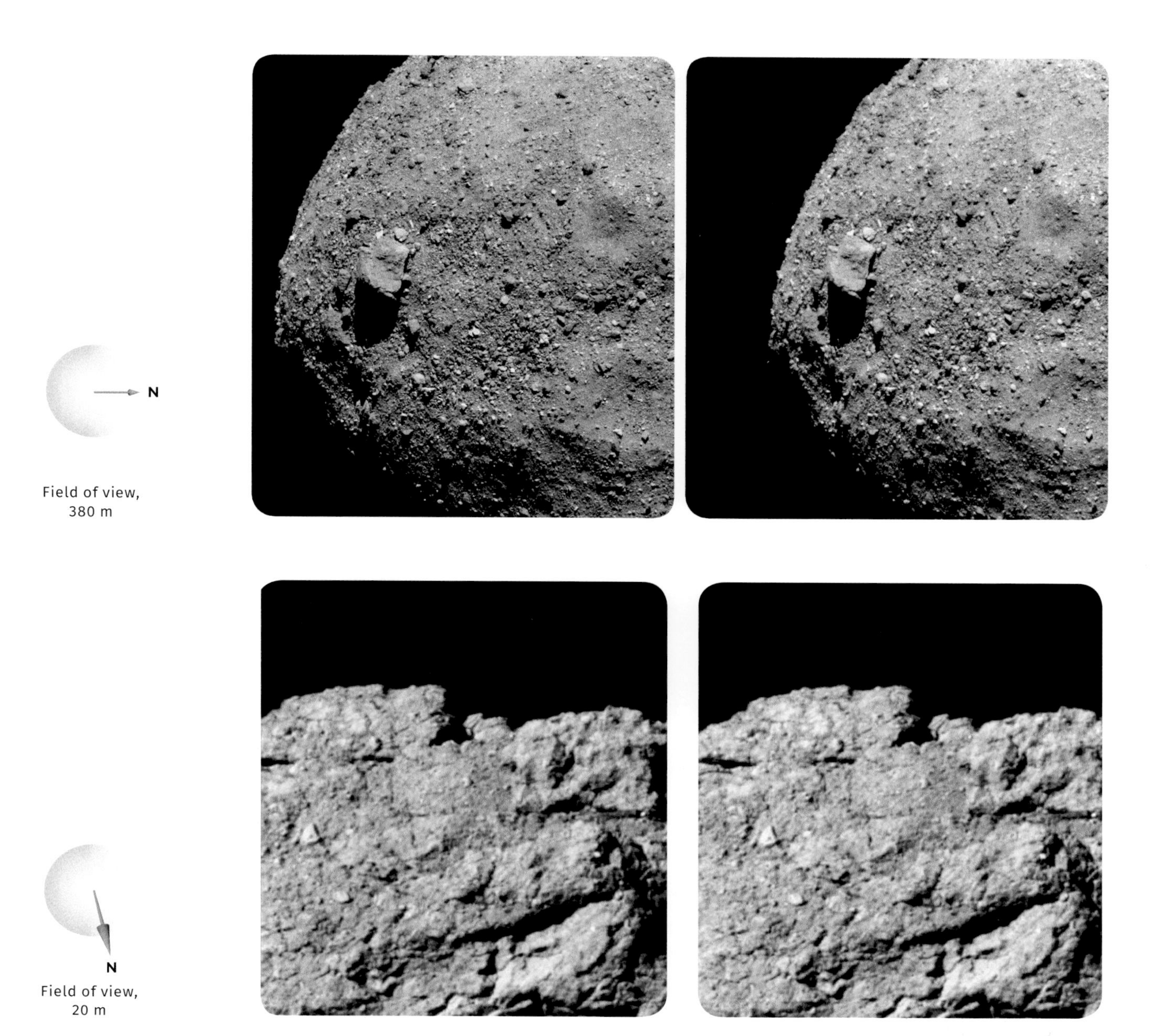
N
Field of view,
380 m
N
Field of view,
20 m

Huhuk Crater

Huhuk Crater is named for the huge fork-tailed bird from Native American Pawnee lore that creates thunder by beating its wings. With a diameter of more than 80 m (260 ft), it is one of the largest impact craters on Bennu. Like Alicanto and Ohnivak, Huhuk Crater is located on the equator and has a pronounced rim, indicating that the surface was deformed by the impact. These large craters are inferred to have formed while Bennu was still in the main asteroid belt, because it has not been in near-Earth space long enough to establish the observed crater population.

Field of view,
170 m

Pouakai Saxum

Pouakai Saxum is named after a monstrous bird of Māori legend who kills and eats humans. Close to Pouakai is a smaller rock whose striking brightness relative to the rest of the surface made it stand out in images taken during the spacecraft's approach. This gave OSIRIS-REx scientists an early hint that exogenic material — originating from other asteroids — is present on the surface (Chapter 7). The bright rock is too small to receive an official name, so Pouakai Saxum was named to provide a close-by landmark.

Field of view, 25 m

Boobrie Saxum

Boobrie Saxum is a quintessential example of a large, dark boulder on Bennu, except that it is located in the northern hemisphere, whereas most other boulders of this scale reside in the south. It has a rough, cauliflower-like texture that includes observable clasts and bright inclusions. Boobrie is a mythological shape-shifter, often taking the form of a giant water bird imagined to inhabit lakes in western Scotland.

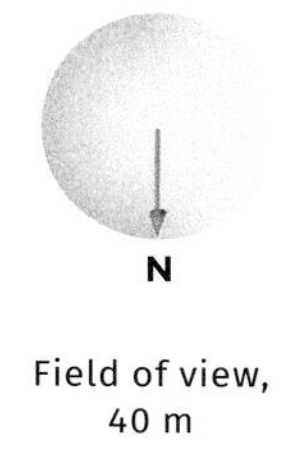

Field of view, 40 m

Ciinkwia Saxum

Ciinkwia Saxum has a flaky texture that suggests the material is weak and friable. In Algonquian lore (in particular the Miami Nation of central North America), Ciinkwia are Thunder Beings, giant eagles who cause thunder and lightning.

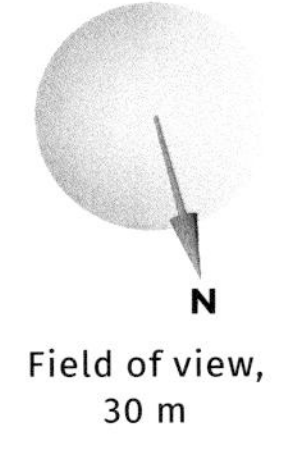

Field of view, 30 m

Amihan Saxum

Amihan Saxum is named for the Tagalog (Philippines) deity depicted as a bird who first inhabited the universe. This boulder is likely larger than its apparent 30 m (100 ft) extent — the bulk of its volume may be buried beneath Bennu's rugged surface. In the false-color image that opens this chapter, this boulder shows a shift from a redder spectral slope on the partially buried southern side to a bluer northern side. This color gradation is thought to result from the equator-facing (in this case, northern) side being more exposed to space weathering effects such as meteoroid bombardment and the Sun's heat.

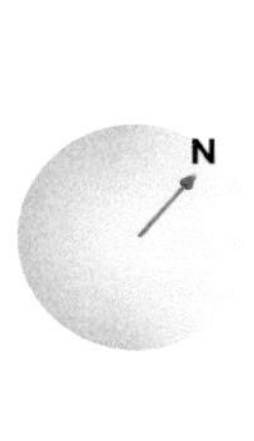

Field of view, 70 m

N
Field of view,
260 m

Tlanuwa Regio

In Native American Cherokee legend, Tlanuwa are giant birds of prey who killed a serpent and scattered the ground with its pieces, which turned into standing pillars of rock.

Affectionately dubbed “Boulder Town” by OSIRIS-REx scientists, Tlanuwa Regio, Bennu’s only named region to date, has the highest concentration of boulders on the asteroid’s surface. This 300 m (1,000 ft) long cluster of boulders could be indicative of a slope collapse and subsequent landslide. Five of Bennu's named boulders are contained within this region (Ciinkwia, Camulatz, Amihan, Odette, and Odile; see the map on page 70).

Camulatz Saxum

Camulatz Saxum is named after a large bird that bit the heads off the first race of people in the K'iche' Mayan (Guatemala) creation myth. This rock appears to have a smoother texture than many of the other dark boulders on Bennu. At the same time, it exhibits phase reddening (an increase in spectral slope with phase angle, Chapter 7), indicative of roughness or dust coating on its surface.

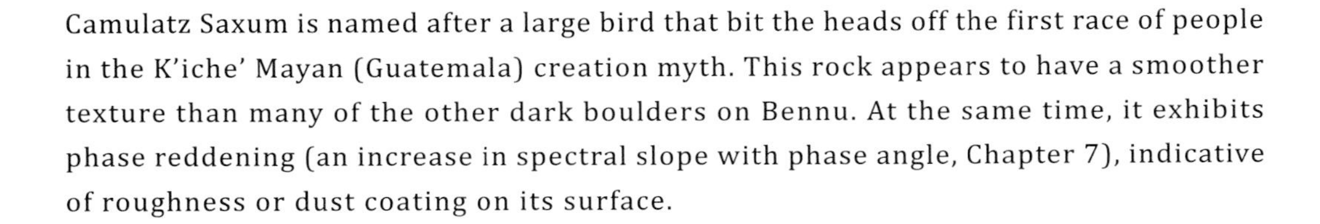

N

Field of view, 25 m

Minokawa Crater

Minokawa, dwarfing the other named craters with its diameter of more than 180 m (590 ft), is located on Bennu's equator. This crater has been almost obliterated by a landslide, the remnants of which can be seen cresting the western rim.

Minokawa is named after a massive bird in Bagobo (Philippines) legend who lives beyond the eastern horizon and causes lunar eclipses.

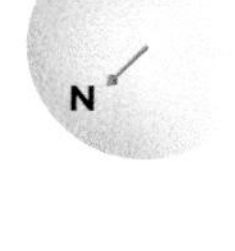

Field of view, 200 m

Odette and Odile Saxa

Odette, the brighter boulder, and Odile, the darker one, were named for the two main characters in Russian composer Pyotr Ilyich Tchaikovsky's ballet *Swan Lake* (1877). These two neighboring boulders epitomize the two main rock types on Bennu: brighter and denser versus darker and more porous.

N

Field of view, 30 m

Bralgah and Dinewan Craters

At 70 m (230 ft) in diameter, Bralgah Crater is the most prominent impact feature in Bennu's southern hemisphere. The diminutive Dinewan Crater (<10 m (33 ft) diameter; shown close up in the smaller mono image) is located inside Bralgah. In the stereo image, Dinewan is near the left (north) edge of Bralgah. A relatively smooth area extends to the north of the two craters (upward in the full-page image), interpreted as a blanket of ejecta created by the Bralgah-forming impact.

In Aborginal myth (Murray River basin, Australia), Bralgah is a crane who tossed an egg from the nest of the emu Dinewan into the sky, where it became the Sun.

N

Field of view,
210 m

Ocypete, Aellopus, and Celaeno Saxa with Pegasus Crater

Pegasus Crater (~10 m (33 ft) diameter) is named for the winged horse in Greek mythology. It has dark fine-grained material at its center, which initially made it a candidate sample collection site (Chapter 5). However, the lighter, squarish Ocypete Saxum (shown close up on the right page, top) and its neighboring, unnamed dark triangular boulder, represented spacecraft hazards. In addition, the eastern crater wall was judged too steep for spacecraft stability.

Aellopus and Celaeno Saxa (right page, middle and bottom) are nearby boulders of similar scale to Ocypete. This area is the source region of the first observed particle ejection event on Bennu (Chapter 3). Because of their association with particles apparently blasting away from Bennu, this trio of boulders was named for the three harpies in Greek mythology, half-woman and half-bird personifications of storm winds.

N

Field of view, 20 m

This stereo view shows the Pegasus Crater and Ocypete Saxum.

All four features are indicated in this oblique view (clockwise from top right: Aellopus, Celaeno, Pegasus, Ocypete).

Ocypete.

Aellopus.

Celaeno.

Computer-generated visualization of the OSIRIS-REx spacecraft approaching the chosen sample site.

Chapter 5

Search for the Sample Site

Before OSIRIS-REx began its journey to Bennu, two other near-Earth asteroids had been the targets of spacecraft rendezvous missions: Eros and Itokawa (Chapter 1). These asteroids have large boulders on their surfaces, but they also have substantial swaths of fine-grained material, like a sandy beach or the fine regolith on the Moon. Thermal and radar measurements of Bennu taken by telescopes indicated that OSIRIS-REx would find similarly open, sandy areas upon arrival, and the mission was therefore designed to touch down in a region with at least 25 meters (82 feet) of clearance in every direction.

However, as the preceding chapters illustrated, boulders are much too prevalent on Bennu's surface for this to be possible. Fine-grained material — far from being available in wide swaths — is concentrated in small, difficult-to-access craters. This terrain made it an unexpected challenge to identify a site for sample collection. To tackle this problem, the team needed a systematic approach to finding locations where the Touch-and-Go Sample Acquisition Mechanism (TAGSAM) could successfully operate. Two major qualitative descriptors of the surface ultimately drove this search: safety and sampleability.

Computer-generated image of TAGSAM on the surface of Bennu.
The baseplate of TAGSAM is just over 30 cm (11.8 in) in diameter.

Safety was of utmost importance. If OSIRIS-REx were to attempt its sampling maneuver near a boulder taller than the TAGSAM arm, a collision could occur, potentially damaging or destroying the solar panels or spacecraft body. As the science team scoured images for potential sample sites, the navigation team set to work on the problem of safely steering the spacecraft into a much more confined area than had been planned.

Sampleability is the amount of material that TAGSAM can collect at a given location, estimated from the observed surface. This parameter combines the abundance of regolith particles small enough to fit into the TAGSAM head (up to two centimeters (0.8 inches) in diameter) with potential negative effects from rocks large enough to inhibit collection. For instance, rocks wider than 20 centimeters (7.9 inches) could fully block the TAGSAM opening, preventing fine particles from entering. A rock taller than six centimeters (2.4 inches) could tilt the TAGSAM head and create a gap through which most of the compressed nitrogen gas would escape — blowing regolith away from TAGSAM rather than into it.

To further complicate matters, the maximum ingestible particle size of two centimeters (0.8 inches) was smaller than the pixel resolution of most of the images taken by the spacecraft, meaning the sought-after fine particles were essentially invisible. It was therefore necessary to assume that unresolved areas in the images contained sampleable material.

In the spirit of exploring all possible paths to sampling, the mission began with more than 50 candidate sites identified by a variety of methods, including manual inspection of thousands of images with the help of citizen scientists, as well as an automated machine-learning approach. This initial batch of sites was quickly narrowed down on the basis of safety and sampleability to a short list of 16 locations. The team then methodically reduced the number of candidate sites to four, and ultimately two: the primary sample site with the best overall chance of success, and a backup site. A backup was needed because if the sampling maneuver failed at or was aborted near the primary site, the spacecraft's thrusters could cause enough surface disturbance during retreat to preclude a second attempt at the same location.

This chapter walks through the key imagery from the search for a sample site, starting from the top 16. The OSIRIS-REx team relied heavily on stereo images produced by authors Brian May and Claudia Manzoni to explore the surface and verify impressions from mono images during this critical part of the mission.

Candidate locations for sampling, overlaid on a global basemap of Bennu's surface (north is to the right). Numbered sites were selected as the top 16 from a pool of more than 50 candidates. The final four sites (color) were nicknamed for birds found in Egypt.

Honorable mentions

Field of view,
20 m

Site 16. This crater in the southern hemisphere, later named Sampati, was selected as one of the top 16 candidate sites because it has a lower density of large boulders than its surroundings. It was ultimately ruled out owing to its small size (less than 10 m (33 ft) wide) and the presence of rocks too large for TAGSAM to ingest.

N

Field of view,
10 m

Site 15. OSIRIS-REx scientists were drawn to this dark crater because of its two unresolved, and thus presumably fine-grained, patches of terrain. But the boulders scattered in and around this small space posed a safety concern.

N

Field of view,
25 m

Site 14. Looking carefully at this stereo image, three smooth areas with sampling potential can be seen: one in the upper center (to the left of the large, dark boulder at the top of the image); another to the right of the same boulder (partially cut off by the image edge); and a third at the bottom (to the right of the lighter-toned large boulder). However, the site was littered with hazardous boulders larger than 20 cm (7.9 in).

N

Field of view,
15 m

Site 13. This site is inside a small crater next to a large, angular boulder (upper middle of the image). The smooth patch in the base of the crater is only 2 m (6.6 ft) wide, and although the unresolved terrain hints at sampleable material, the small size of the site and its proximity to the boulder made it unsuitable.

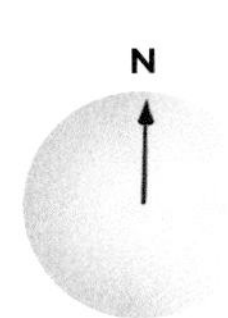

Field of view,
15 m

Site 12. The patch of unresolved terrain in the center of this image suggests fine sampleable material, but is only ~3 m (9.8 ft) in diameter. In addition, the brightness variation across the surface suggests that particles larger than 10 cm (3.9 in) are present, which could tilt the TAGSAM head and impede sampling.

N

Field of view,
30 m

Site 11. This site is the only candidate that sits atop what appears to be a large, mostly buried boulder. This location was selected under the rationale that it could serve as a safe, relatively flat place to collect dust particles in TAGSAM's contact pads — a worst-case scenario if safe sites with better sampleability could not be found.

N

Field of view, 40 m

Site 10. This candidate site is a small crater in the relatively smooth basin of Bralgah Crater (Chapter 4). It was eliminated based on comparison with a more suitable nearby candidate (Site 3).

N

Field of view, 25 m

Site 9. Also located in Bralgah Crater, this site consists of patches of unresolved terrain along the southern crater wall. Though fine-grained material is present across a large area, the even distribution of hazardous rocks eliminated this site from consideration.

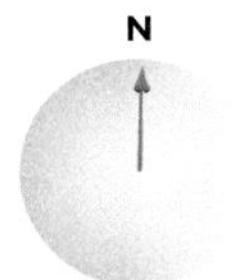

Field of view,
25 m

Site 8. This site is located at the base of a crater with a large, distinctive boulder on its southern edge. Smaller boulders trace the rim of the crater — a common occurrence on Bennu, reminiscent of campfire rings. This site was a strong contender but was eliminated in favor of the nearby, more amenable, Site 1.

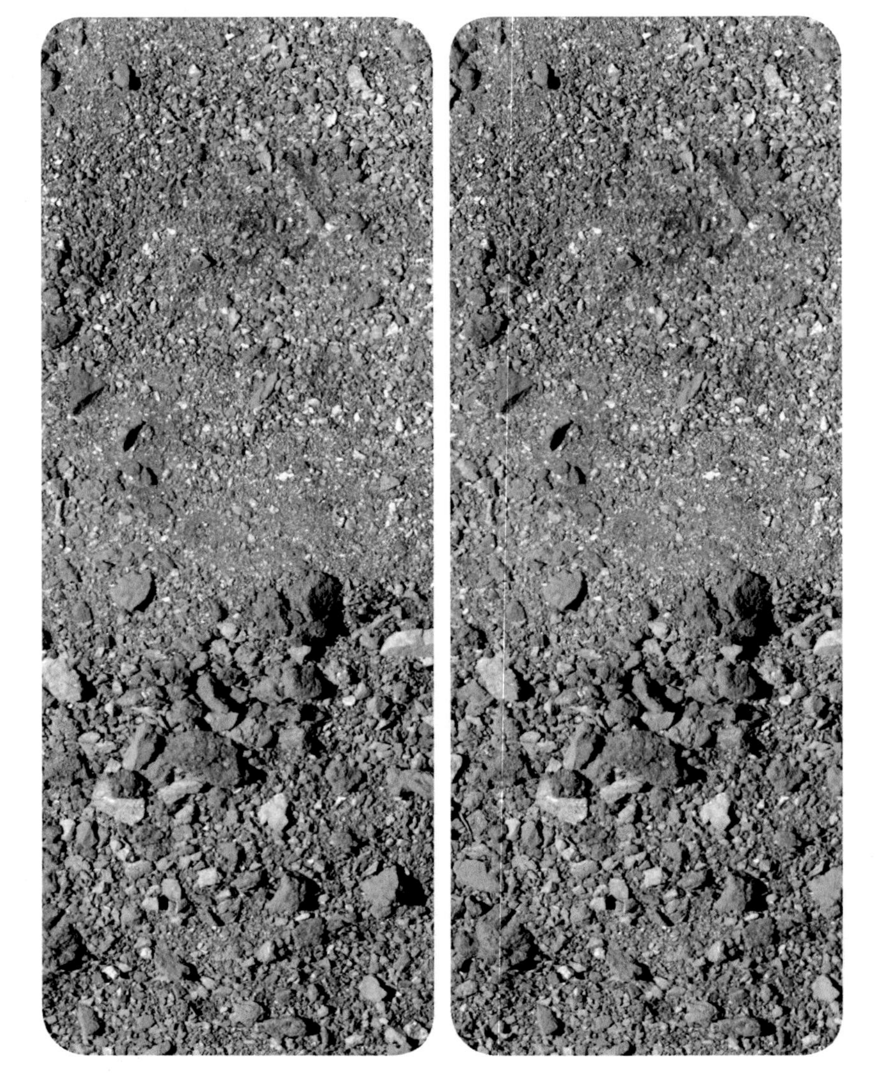

N

Field of view,
30 m

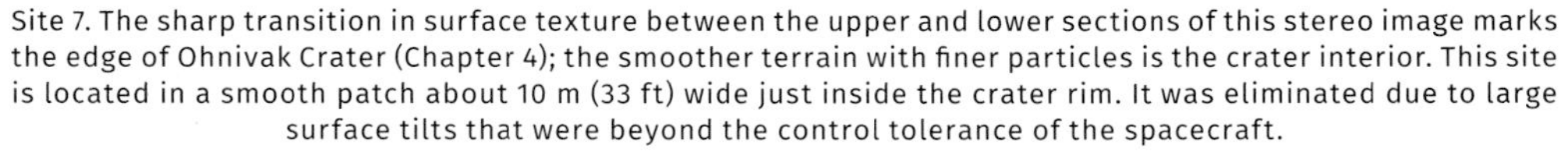

Site 7. The sharp transition in surface texture between the upper and lower sections of this stereo image marks the edge of Ohnivak Crater (Chapter 4); the smoother terrain with finer particles is the crater interior. This site is located in a smooth patch about 10 m (33 ft) wide just inside the crater rim. It was eliminated due to large surface tilts that were beyond the control tolerance of the spacecraft.

N

Field of view,
25 m

Site 6. This site is centered on a dark crater that is surrounded by smaller impact scars. The team investigated each unresolved patch in hopes of finding a suitable touchdown location. In the end, the scattering of potentially TAGSAM-blocking boulders over this area made it less than ideal.

N

Field of view,
20 m

Site 5. This site showed promise with what appeared to be an abundance of fine-grained material in Pegasus Crater (Chapter 4). It was ultimately passed over because of safety concerns posed by the crater's deep yet narrow shape, revealed in the stereo image, and the two boulders on its western rim (Ocypete Saxum and its unnamed companion).

The final four

After an exhaustive review of the global surveys of Bennu, mission scientists and engineers winnowed the field of potential sample sites down to the four shown here. These sites were discussed so much that the team gave them nicknames, each for a bird found in Egypt, in keeping with the Egyptian and avian influences on the mission's lexicon.

Site 4, nickname Kingfisher. The center of this 7-m (23-ft) diameter crater is free of large rocks and appears to contain sampleable material. Its tight borders, however, posed challenges to navigation.

N

Field of view, 25 m

Site 3, nickname Sandpiper. This site is located in an appealingly flat area of Bralgah Crater's southern wall. Particle motion down the crater wall has created a long patch of fine-grained regolith trapped between larger rocks. The generally flat nature of this terrain gave it a high safety score, and it was a strong contender for the final two.

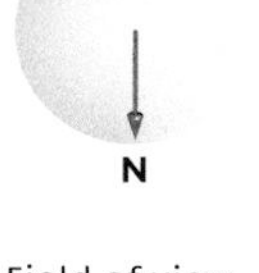

Field of view, 15 m

N
Field of view,
25 m

Site 2, nickname Osprey. Located in Wuchowsen Crater (Chapter 4), this site offered several regions where fine particles were concentrated, and the lack of nearby boulders gave it the highest safety score. However, the team debated whether the dark area in the center consisted of fine particles or the protruding face of a buried boulder. Another cause for concern was the abundance of particles that may have been too large to ingest, giving it relatively low sampleability.

N
Field of view,
30 m

Site 1, nickname Nightingale. The northern location of this site in Hokioi Crater (Chapter 4), and the imposing boulder on the eastern rim — known to the team as Mount Doom — presented a more challenging navigation problem than Osprey. But Nightingale had the greatest abundance of unobstructed sampleable material in a space large enough for the spacecraft to access, making it a highly desirable target.

A closer look

With these four sites in mind, the team embarked on the Reconnaissance flyovers (Chapter 2) to acquire close-range, localized observations that would yield high-resolution image mosaics. The spacecraft scanned a larger area of the surface than necessary to ensure that the target sites would be imaged even if the spacecraft went slightly off track. The resulting images, with a pixel scale of just 1 centimeter, revealed Bennu's surface in the unprecedented detail needed to downselect from four to two candidate sites.

Kingfisher: site diameter 8 m (25 ft).

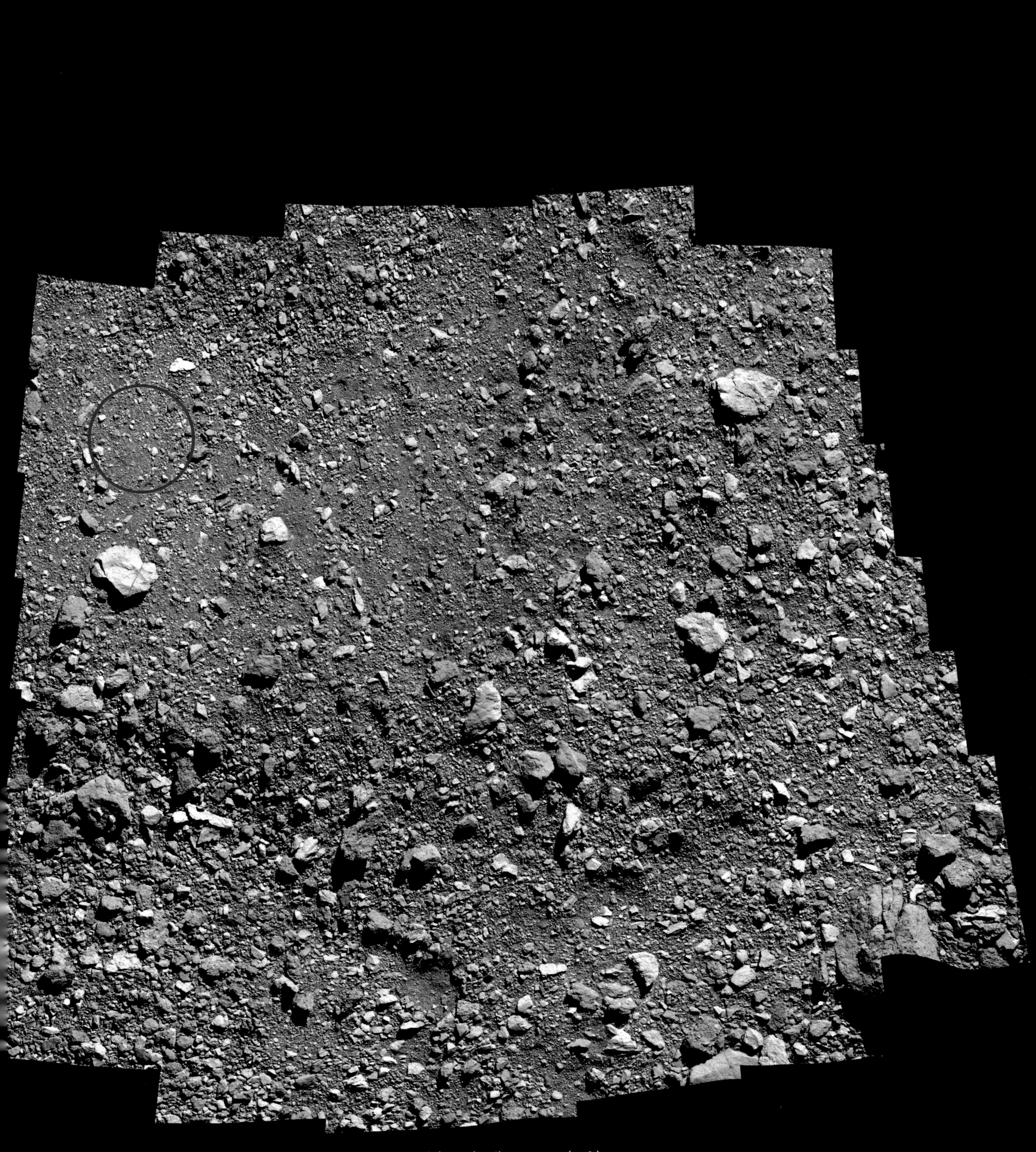

Sandpiper: site diameter 8 m (25 ft).

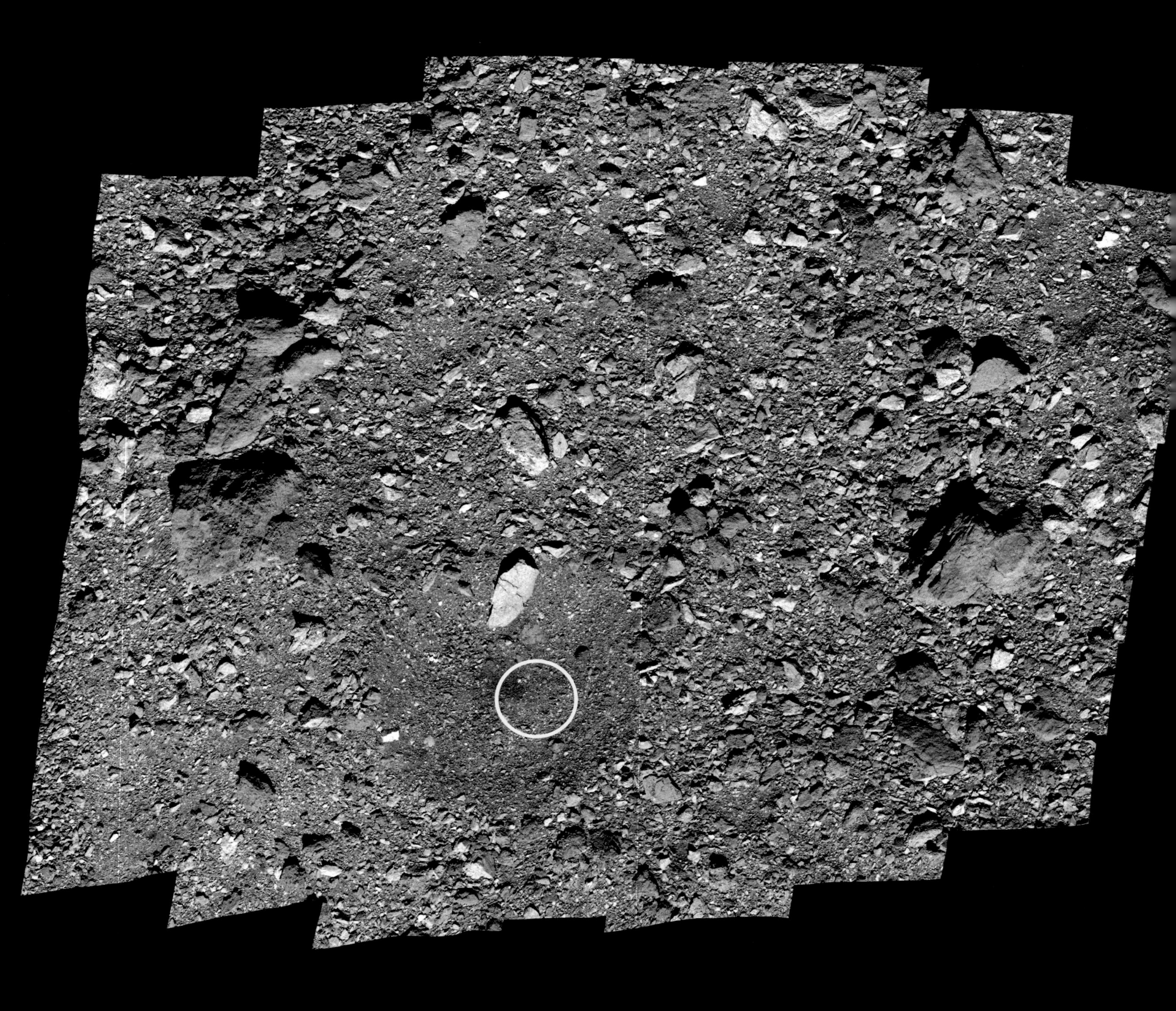

Osprey: site diameter 6 m (20 ft).

Nightingale: site diameter 8 m (25 ft).

The final two

Eliminating Kingfisher was fairly straightforward because although it had similar surface characteristics to Nightingale, it was much smaller. Sandpiper was initially favored based on a high likelihood of making safe contact with the surface, but ultimately the safety at Osprey proved even higher. The team therefore selected Osprey and Nightingale as the final two sites, but the question remained of which would be primary and which backup.

Meanwhile, as mission engineers improved OSIRIS-REx's autonomous navigation system, confidence grew in the precision of the touchdown location. The sampling maneuver that once needed 25 meters (82 feet) of margin in radius now could be done, astonishingly, with less than five meters. In the following images, the yellow and blue circles respectively outline the final borders of the Osprey and Nightingale sites, within which the navigators had high confidence in their ability to deliver the spacecraft. Over these areas, the team mapped surface factors pertinent to safety and sampleability to directly compare the two sites.

Choosing between Nightingale and Osprey for the primary versus backup site was not easy. It would be less challenging to navigate the spacecraft into Osprey, but Nightingale offered more pockets of sampleable material. The thematic maps shown here are just a subset of those that helped differentiate between the two sites and led the team to designate Nightingale as the chosen touchdown location. Its selection as the primary site was finally validated when the TAGSAM safely collected an impressively large sample from its center, as described in the next chapter.

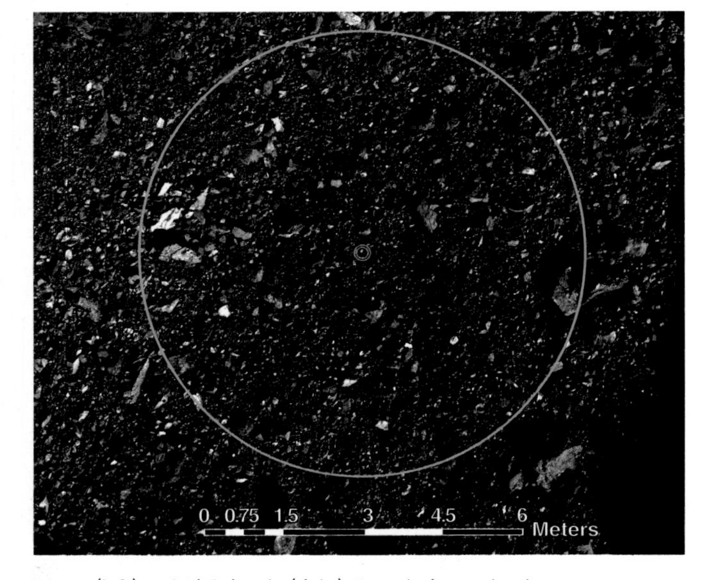

The highest-resolution images obtained of the final two sites: Osprey (left) and Nightingale (right). For relative scale, the circumferences of the TAGSAM baseplate (outer orange circle) and opening (inner orange circle) are shown at the center of each site.

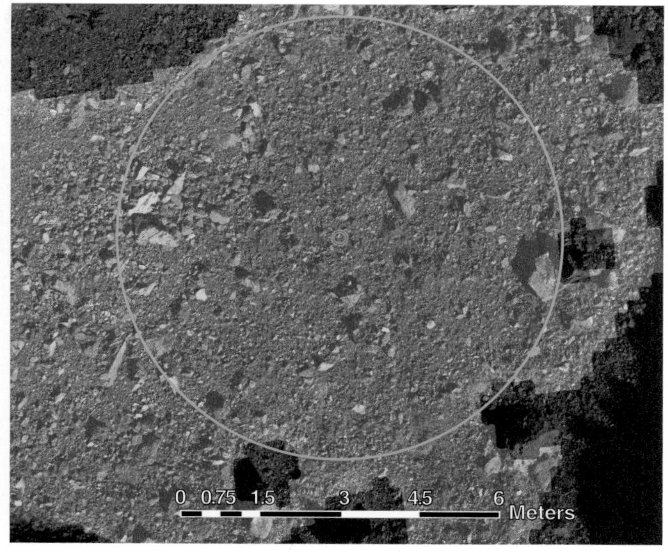

Maps of surface hazards were created using the fine-scale altimetric data acquired by OLA (Chapter 2). Areas shaded green indicate safe areas for the spacecraft to contact, whereas those shaded red would be dangerous. Yellow indicates a margin of avoidance around hazardous objects.

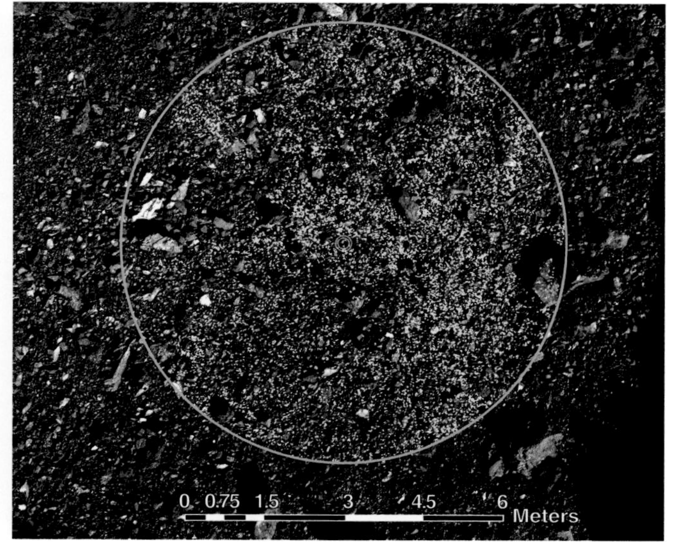

Sampleability was assessed in part by measuring the size of the surface particles. The goal was to compare availability of collectible particles between the two sites. The yellow markers indicate those particles that are 2 cm (0.8 in) or smaller. These measurements were painstakingly made by hand by a team of 8 boulder counters over 60 hours.

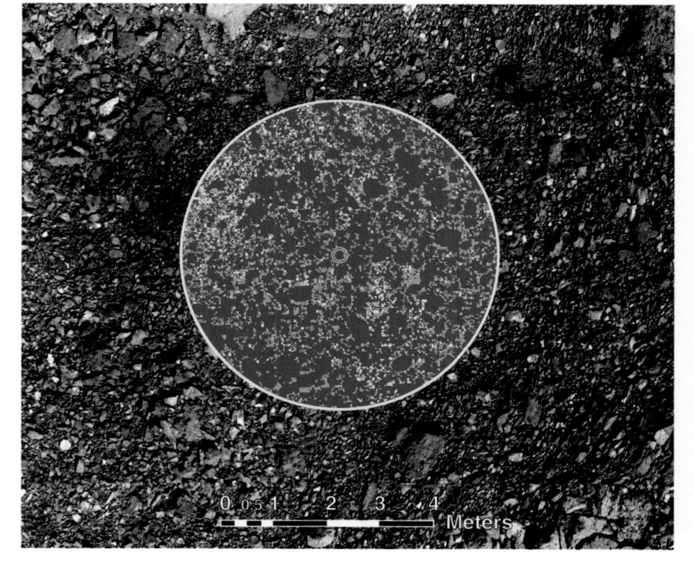

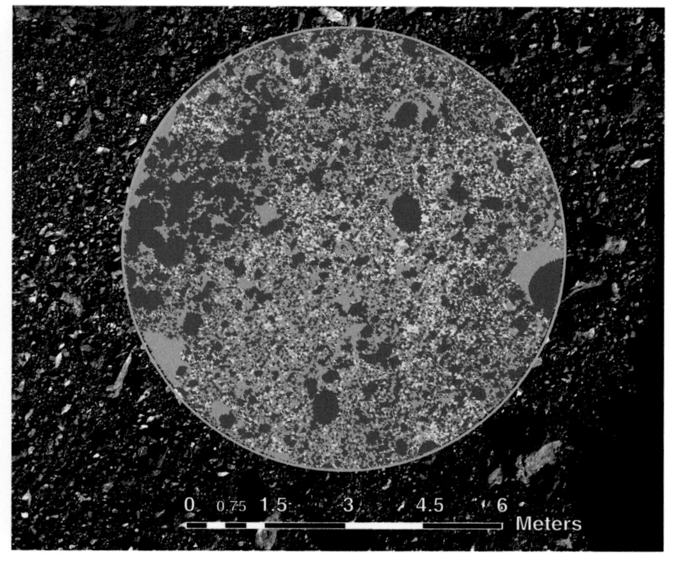

Another way to visualize sampleability is to shade the surface by the size of particles covering it. Here, red represents particles too large to sample, yellow represents particles confirmed to be sampleable, and blue represents unresolved areas that presumably include fine, sampleable particles.

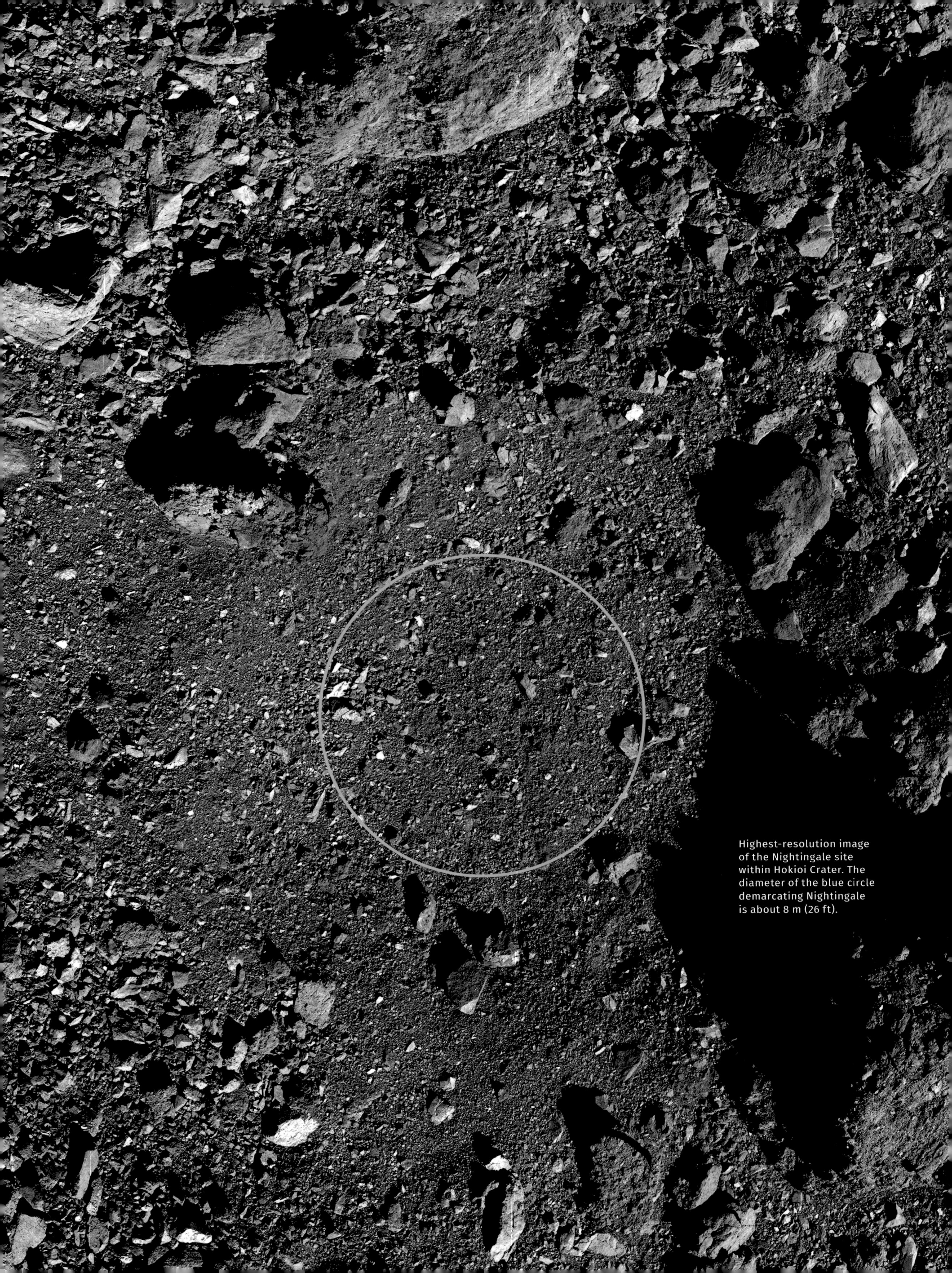

Highest-resolution image of the Nightingale site within Hokioi Crater. The diameter of the blue circle demarcating Nightingale is about 8 m (26 ft).

Chapter 6

Touch and Go

Finally, more than two years after the spacecraft captured its first image of Bennu, OSIRIS-REx's historic moment arrived. On October 20, 2020, the spacecraft targeted site Nightingale to collect a sample of rocks and dust. Nightingale had been selected as the mission's primary sample site because it held the greatest amount of unobstructed fine-grained material and met the mission safety criteria (Chapter 5). The Touch-and-Go (TAG) sampling maneuver, however, the culmination of years of effort, would require a complex sequence of events.

During sampling, the team needed to ensure that the spacecraft, which is the size of a large van, would safely touch down in a narrow area that was only the size of a few parking spaces. Even though it was deemed safe, the site was surrounded by building-sized boulders, which represented potential spacecraft hazards. Having assessed the difficulty of the terrain, the team decided to upgrade the spacecraft with an onboard navigation system including a hazard map of site Nightingale, which delineated objects, such as meter-scale boulders, that could damage the spacecraft. Because Bennu was over 300 million kilometers (186 million miles) away, there was an additional challenge in that there was an 18-minute delay each way in communication between the spacecraft and mission control. Thus, for the final sequence, the spacecraft was on its own, and sampling operations were fully autonomous.

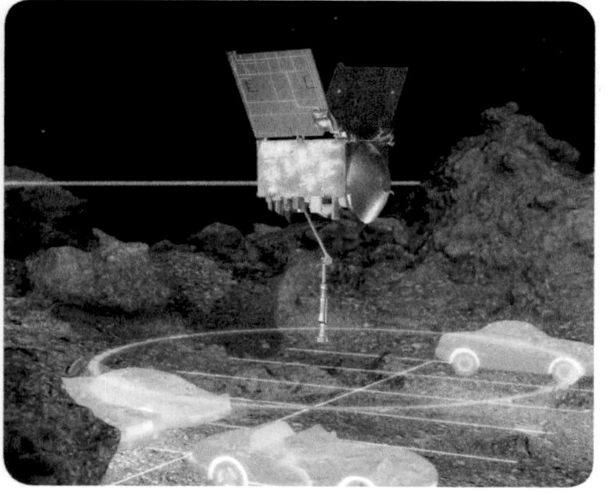

Stereo image illustrating the size of the Nightingale site relative to a parking lot.

Descent

The descent sequence began when OSIRIS-REx fired its thrusters for orbit departure 700 meters (2,300 feet) above Bennu's surface. Heading towards site Nightingale, four hours later, the spacecraft performed the Checkpoint maneuver at an altitude of 125 meters (410 feet) (Chapter 2). This burn adjusted the spacecraft's position and its speed toward the surface. The following series shows approximately concurrent NavCam (left) and SamCam (right) images during the final descent to Nightingale and through the sample collection event. SamCam was pointed straight down at the surface, with TAGSAM filling the central field of view. NavCam was off-pointed, looking to the side of spacecraft. The cameras' fields of view overlapped when the spacecraft was at higher altitudes.

About 11 minutes after Checkpoint, the spacecraft performed the Matchpoint burn at 55 meters (180 feet) altitude. This maneuver slowed the spacecraft's descent and matched the surface motion resulting from the asteroid's rotation. At this point in its journey,

In the NavCam image (left), taken post-Checkpoint, Hokioi Crater is visible in the lower center, with Nightingale circled. Mount Doom appears on the left edge of the SamCam image (right).

The northern rim of Hokioi Crater can be seen in the NavCam image on the left. The SamCam image (right) reveals the generally benign surface of Nightingale, with some large hazardous boulders at the bottom.

the spacecraft was hovering directly above site Nightingale, allowing Bennu's gravity to pull it in for touchdown.

After Matchpoint, the spacecraft's computer analyzed NavCam images to determine if it was on course to touch a hazardous zone. If so, the spacecraft could autonomously wave off its approach. This decision would occur at an altitude of five meters (16 feet), meaning that the thrusters would perturb the surface, rendering the site unusable for a future attempt. This strategy was designed to keep the spacecraft safe and would have allowed for a subsequent sample collection attempt at backup site Osprey.

Once the spacecraft declared the contact location safe, it prepared to touch down. The attitude control thrusters were enabled to protect against rapid tip-over. A timer was initiated to trigger the sampling sequence even if surface contact was not sensed. Fault protection was disabled, meaning that no computer glitch could interfere with the sample collection event. Once contact was sensed, TAGSAM was ready to fire its pressurized nitrogen gas to agitate the surface and direct regolith particles into the collection chamber within the head.

Computer-generated image of the spacecraft during its final descent into Hokioi Crater.

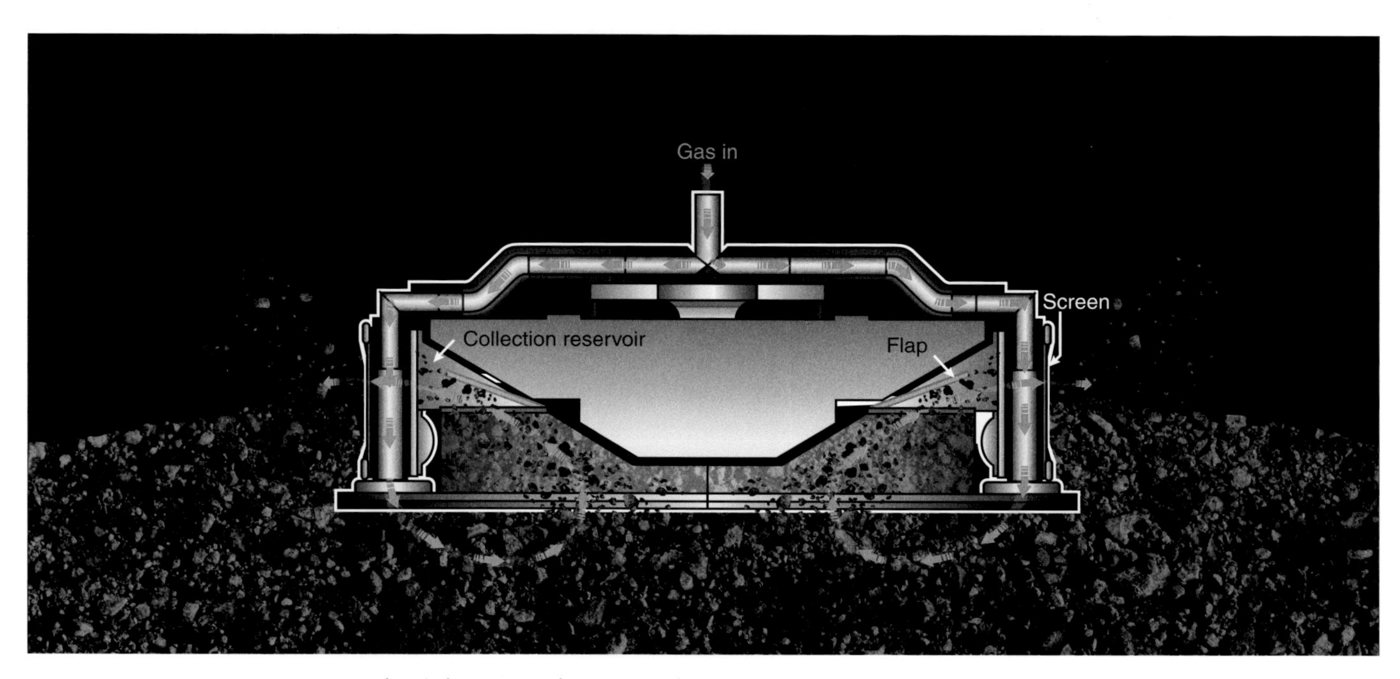

Cross-sectional diagram showing how the nitrogen gas flows through TAGSAM to drive loose regolith into the collection chamber. The arrows indicate the direction of flow.

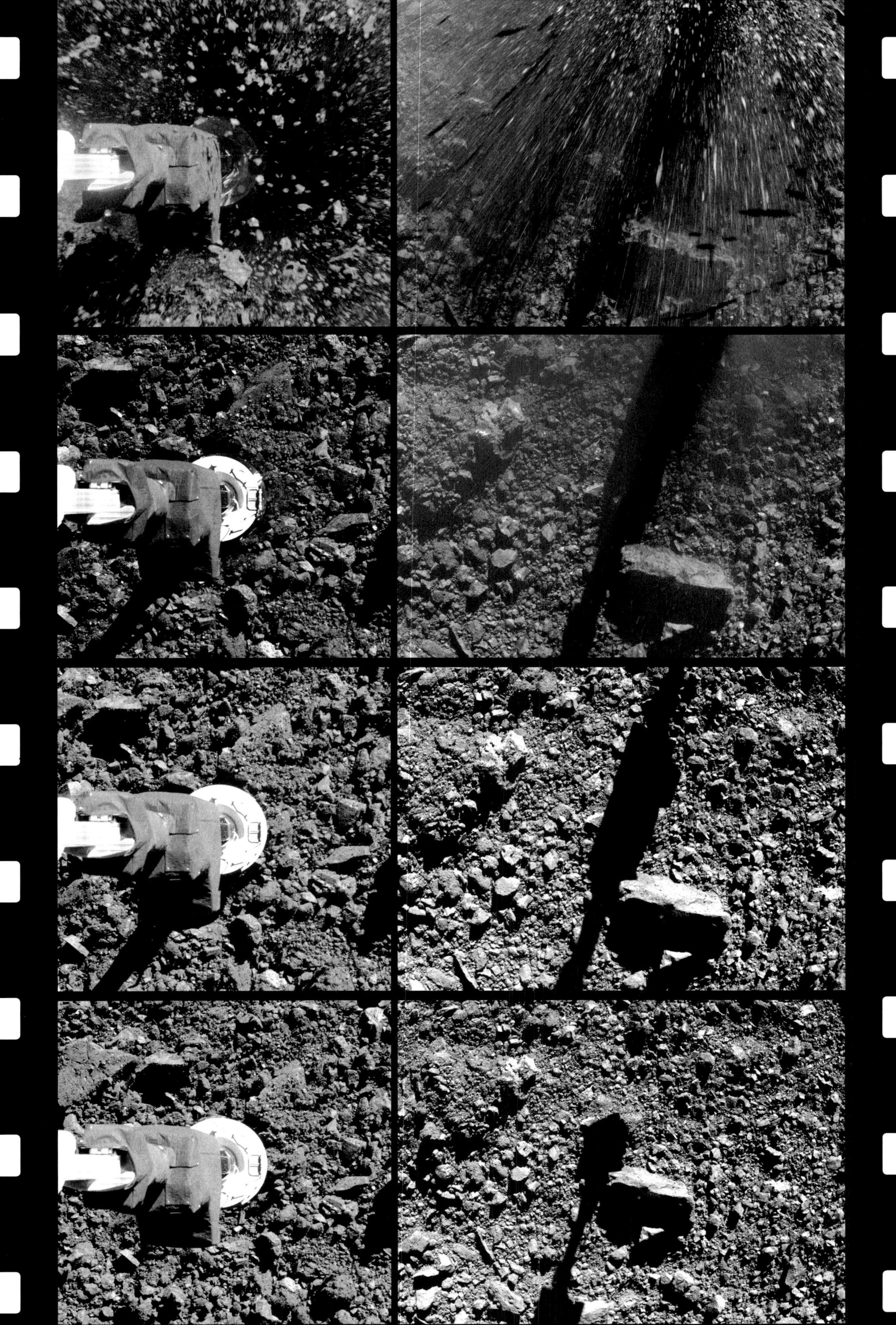

Contact

TAGSAM contacted the surface with a vertical velocity of just ten centimeters (four inches) per second, about the speed of a baby crawling. The TAGSAM head appeared to crush some of the porous rocks underneath it. One second after contact, after penetrating ten centimeters into Bennu's unexpectedly soft surface, TAGSAM fired its burst of nitrogen gas, sending up a chaotic cloud of dust and rocks.

On the day of TAG, a small group monitored the spacecraft from Lockheed Martin's Mission Support Area. The spacecraft dribbled back data during the descent at a meager 40 bits per second. Thus, it was a tense wait for confirmation of contact to arrive. Finally, even though wearing the face masks required by the COVID pandemic, the team erupted in cheers of joy at the news of successful touchdown on Bennu. They were joined in celebration by their remote teammates, and, thanks to a livestream, by space enthusiasts all over the world.

Estelle Church, spacecraft chief engineer, declares "It's a touchdown!"

(Opposite) In this sequence of images, TAGSAM's shadow can be seen crossing the NavCam field of view (left). TAGSAM makes contact (third row), sinks into the surface, then discharges its gas (fourth row), excavating regolith and plunging the sample head into shadow.

Backaway and surface disturbance

Six seconds after touchdown, the spacecraft fired its thrusters to back away from Bennu. This maneuver disturbed the surface even more than had been expected. During the backaway maneuver, a giant plume of debris was visible in the NavCam images created from both TAGSAM gas release and the thrusters firing.

Stereo image created by combining NavCam and SamCam images obtained simultaneously during backaway.

A reconstruction of the sampling event, below, illustrates the swirling cloud of material mobilized by the spacecraft's interaction with the unexpectedly soft, compliant surface. It turns out that the particles making up Bennu's exterior are so loosely packed and lightly bound to each other that if a person were to step onto Bennu, they would feel very little resistance, as if stepping into a pit of plastic balls in a child's play area.

Stereo image reconstructing the surface mobilization resulting from sampling.

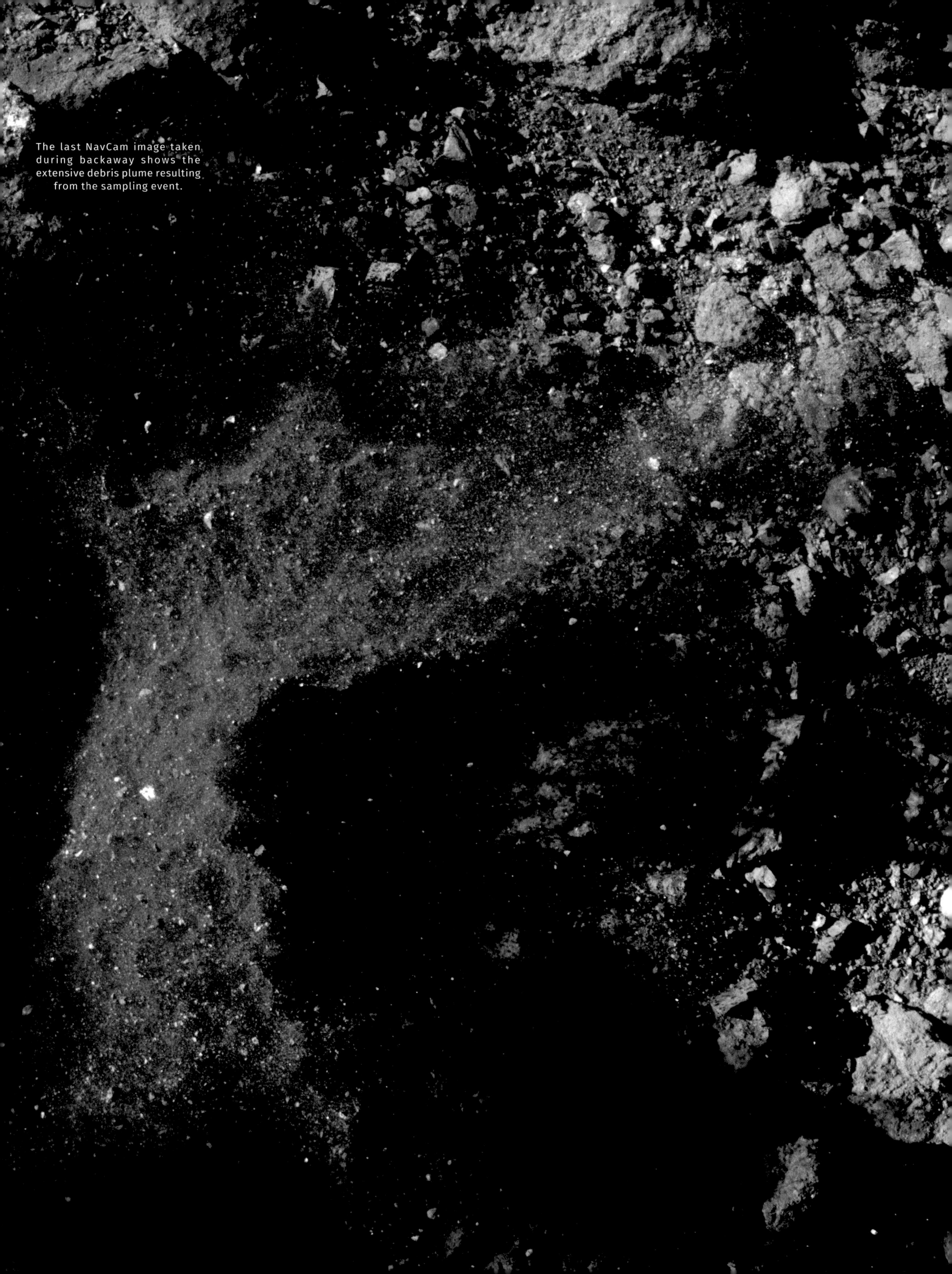

The last NavCam image taken during backaway shows the extensive debris plume resulting from the sampling event.

Nightingale before and after sampling

Immediately after collecting the sample, the spacecraft flew directly away from Bennu in a straight line. The team did not know whether the Bennu environment post-impact would be safe for spacecraft operations. Thus, OSIRIS-REx retreated to a distance of more than 1,000 kilometers (620 miles). Once the debris settled down, the spacecraft returned on April 7, 2021, for a last flyover of Bennu to document surface changes resulting from the TAG maneuver. During the flyby, OSIRIS-REx imaged Bennu for 5.9 hours, covering more than a full rotation of the asteroid. It flew within 3.5 kilometers (2.2 miles) of the surface of Bennu — the closest it had been since the sample collection event.

Comparing the images from the final flyby with those taken two years prior revealed signs of extensive surface disturbance. At the sample collection point, there was a new depression, with several large boulders evident at the bottom, suggesting that they were exposed by sampling. There was an increase in the amount of highly reflective material near the TAG point against the generally dark background of the surface, and many rocks were moved around. Where thrusters had fired against the surface, substantial mass movement was apparent. Multiple sub-meter boulders were mobilized by the thrusters into a campfire ring-like shape — similar to rings of boulders previously seen around small craters pocking Bennu's surface.

The final flyby provided images that were used to create a new 3-D digital terrain model of site Nightingale after sampling. The team had expected to see a modest pockmark the size of the TAGSAM head, about 30 centimeters (12 inches) as a result of the event. Instead, an eight-meter (26 foot) wide elliptical crater was found, confirming that the surface density of Bennu is shockingly low and the cohesion between particles is almost non-existent.

N

Field of view, 25 m

Stereo view compiled from images captured before the TAGSAM event (right frame) and after it (left frame). This pairing was chosen with a suitable baseline to allow the surface to appear in 3-D in our usual slightly hyperstereo, but the area disturbed by TAGSAM has a different appearance in left and right views (upper right). This gives "retinal rivalry", causing confusion in our brains as we try to view the stereo. So, in the interest of not getting a headache here, it may be advisable not to linger too long on the crater area, but concentrate on the edges of the disturbance. This view does not give any indication of the three-dimensional shape of the surface at the impact point, but it does give a very clear visualization of the extent of the disruption.

N

Field of view,
25 m

Nightingale before sample collection. Images acquired during the Detailed Survey in 2019.

N

Field of view,
25 m

Nightingale after sample collection. Images acquired during the final flyby in 2021.

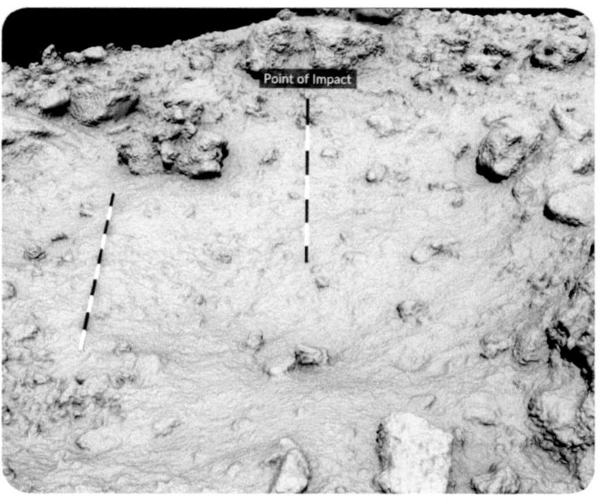

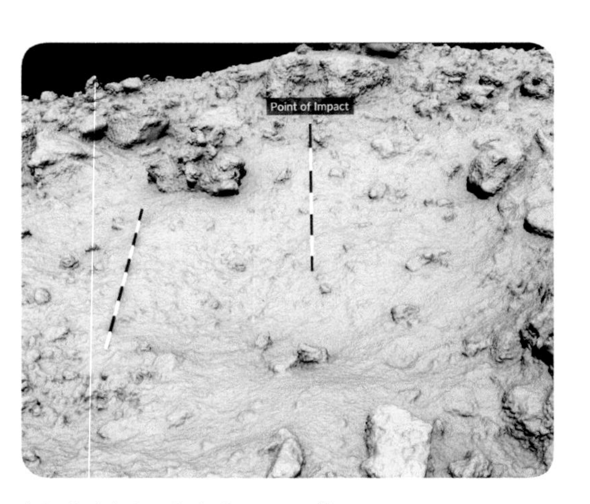

Stereo image of the digital terrain model of Nightingale before sampling.

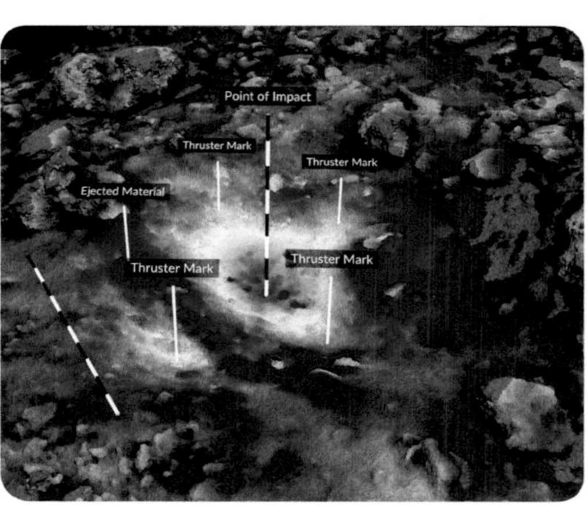

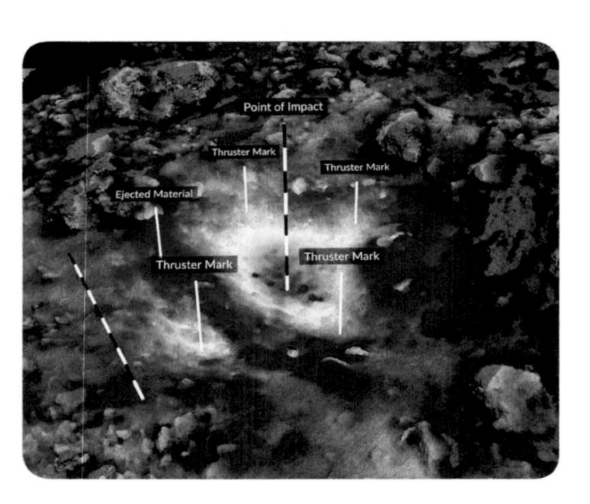

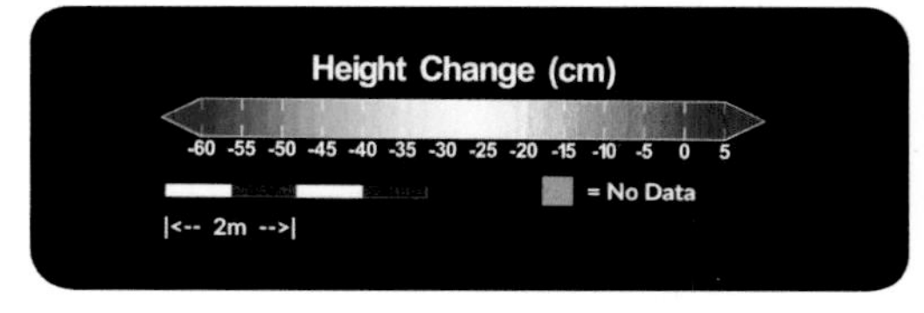

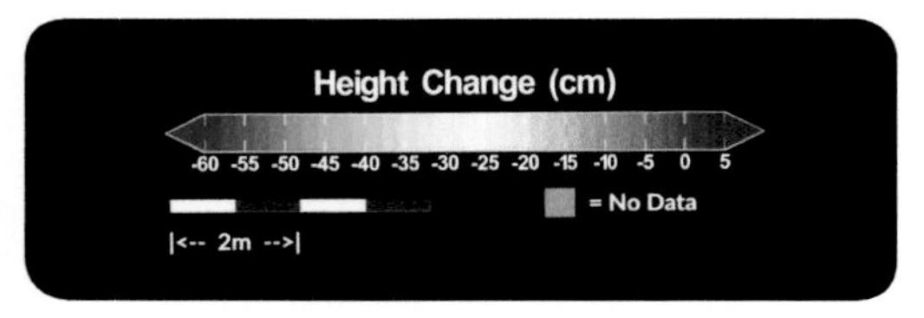

Stereo image of the digital terrain model of Nightingale after sampling, with color coding for the measured changes in surface height.

The sampling event also resulted in noticeable spectral changes. The fresh surface of Nightingale was darker overall, with reflectance decreased by 5%, and spectrally redder. The team attributed the spectral changes to the exposure of unweathered, organic-rich material.

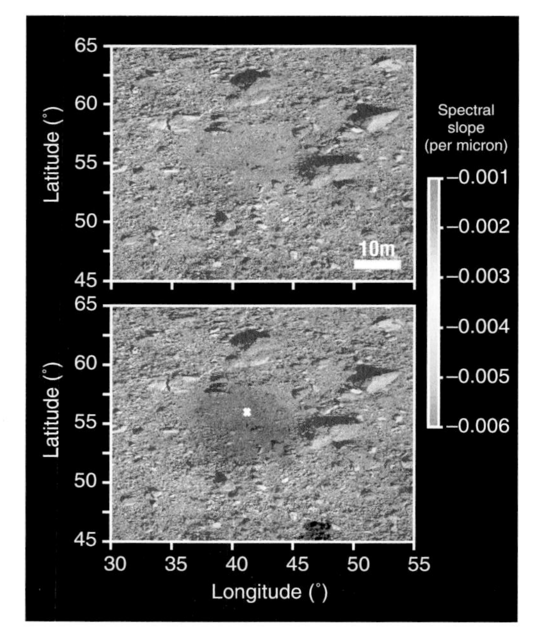

Spectral map of site Nightingale before and after sampling. The freshly exposed material is spectrally "redder" than the original surface. Redness is evaluated using spectral slope, a measure of the brightness as a function of color. The white "X" indicates the TAG point.

Stowing the sample

Before stowing the TAGSAM head, with the sample inside it, in the return capsule, the team needed confirmation that a sample had been acquired. After analyzing images taken during the descent, OSIRIS-REx scientists were able to pinpoint the exact location where TAGSAM made contact. Fortunately, the location was captured by both MapCam and SamCam, providing a perfect stereo image of the collection site. These images revealed that there were more than 35 particles of sampleable size, less than two centimeters (0.8 inches) across, that were directly under the TAGSAM head at contact. The deep penetration of TAGSAM prior to gas firing implies that these surface particles were likely pushed into the collection chamber by the sheer momentum of contact. The team was now confident that an abundant sample had been collected.

Stereo image of the contact location, created by combining a MapCam and a SamCam image from the final approach to the surface.

After sampling, the spacecraft captured images of the TAGSAM head as it moved through several different positions. In reviewing these images, the team noticed that the head appeared to be full of asteroid particles as expected, but in an unpleasant surprise, some of them were escaping. The team concluded that bits of material were passing through small gaps where a mylar flap — the collector's "lid" — was wedged open by larger rocks.

This image of the empty TAGSAM, shown for comparison with the full TAGSAM (overleaf), was taken in November 2018.

This image was taken two days after sample collection and shows a dark TAGSAM interior full of sample, with hundreds of particles fleeing the collector and escaping into space.

The process to stow the sample, which involved moving the TAGSAM head, required the team's continuous supervision and input over a two-day period. The stowage procedure was performed in incremental steps under the supervision of mission controllers to ensure the safety of the sample. On October 28, 2022, the team sent commands to the spacecraft, instructing it to move the TAGSAM head into position and close the capsule. The sample of Bennu was now safely stored and ready for its journey to Earth. The team estimated that at least 150 grams of sample were contained within TAGSAM, more than twice the mission requirement.

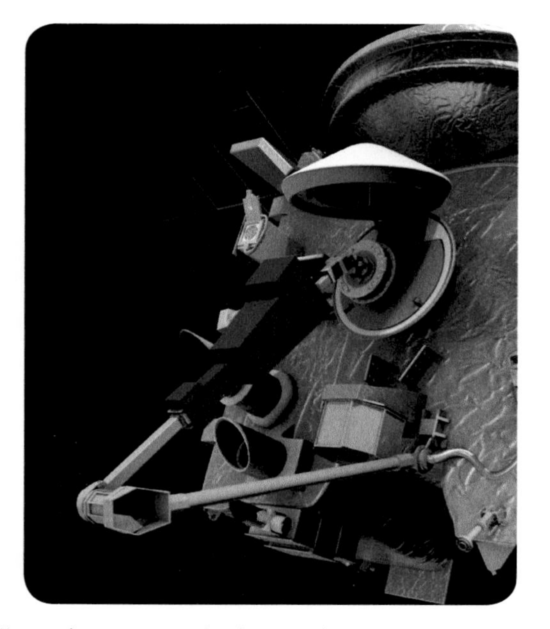

Computer-generated stereo image illustrating TAGSAM using its robotic arm to position the TAGSAM head in the return capsule.

StowCam image of the TAGSAM head hovering over the return capsule, after the TAGSAM arm moved it into the proper position for capture.

The TAGSAM head secured onto the capture ring in the return capsule.

The sealed capsule, with the lid closed and secured.

Departure

On May 10, 2021, the spacecraft fired its main engines full throttle for seven minutes — the most powerful maneuver since it arrived at Bennu in 2018. This burn thrust the spacecraft away from the asteroid at nearly 1,000 kilometers (620 miles) per hour, setting it on a two-and-a-half year cruise towards Earth.

The OSIRIS-REx spacecraft is scheduled to fly over Earth on September 24, 2023, delivering its precious cargo — pristine material from the early Solar System.

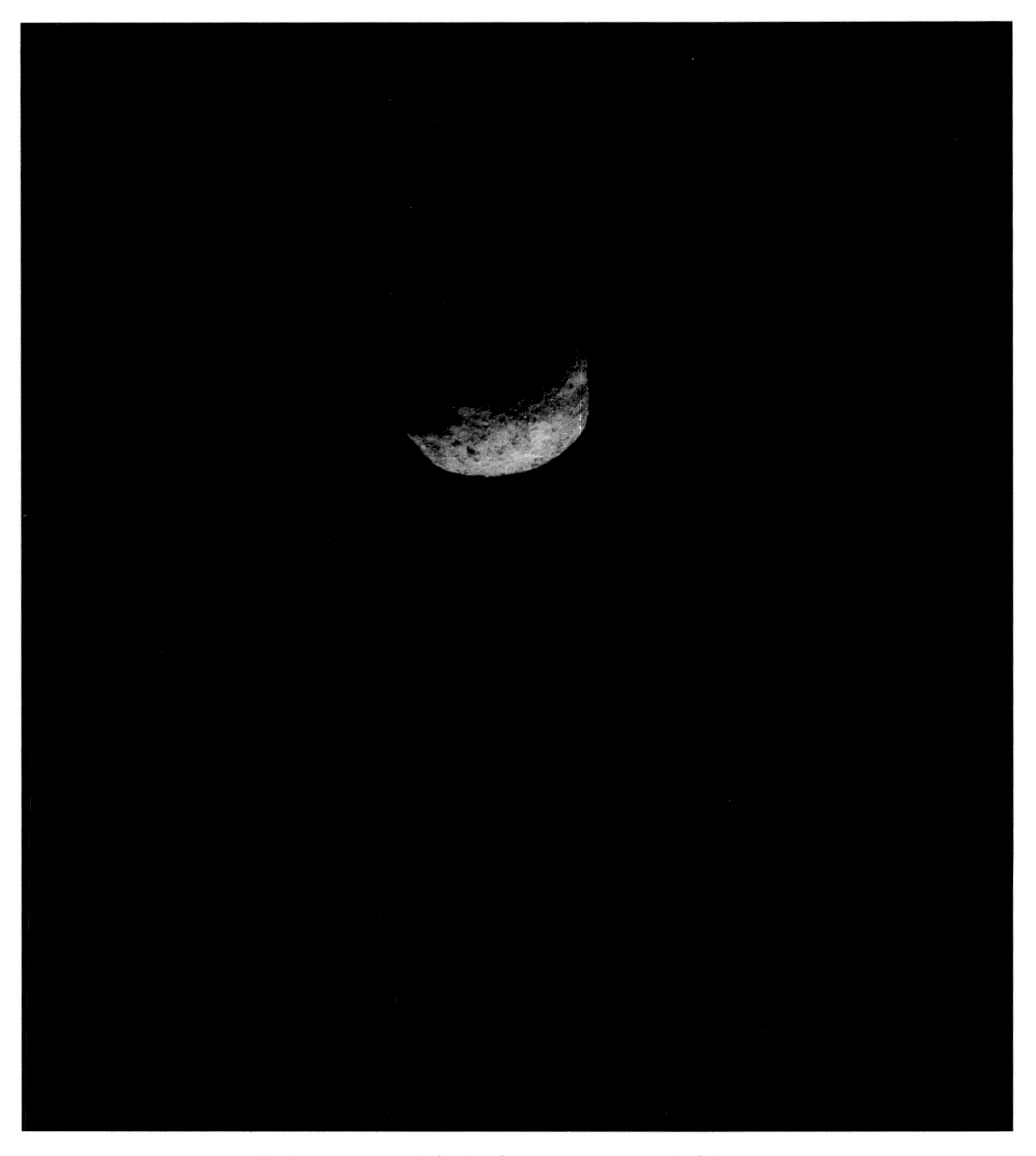

NavCam captured this final image of Bennu on April 9, 2021.

Artist's conception of the spacecraft releasing the return capsule to Earth.

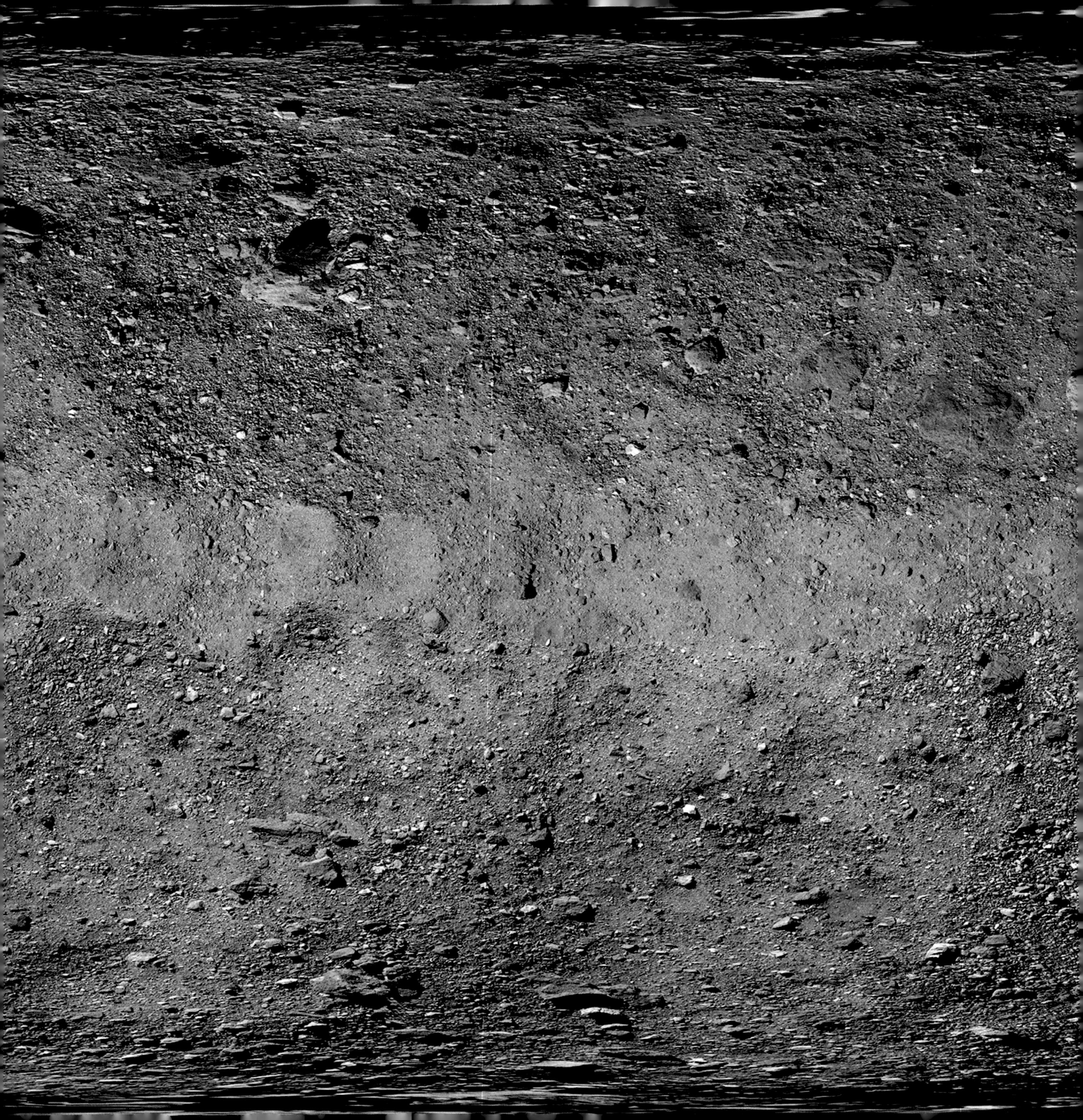

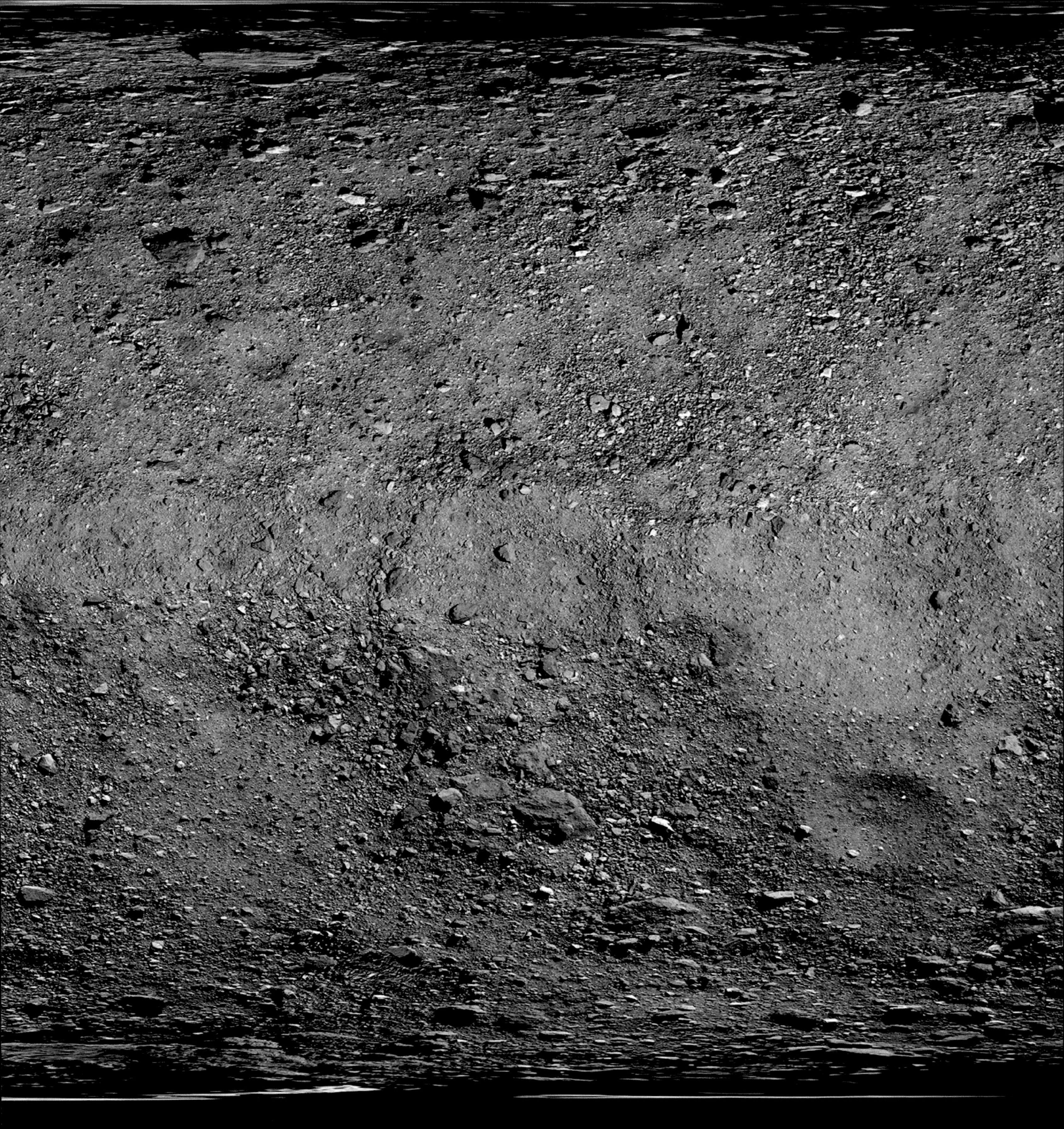

The global basemap of Bennu. In the full-resolution version (too large to print in this space), each pixel represents just 5 centimeters on Bennu's surface.

Still from an animation of the spacecraft mapping Bennu.

Chapter 7

Atlas of Bennu

The key to OSIRIS-REx's sampling success was the mission team's thorough familiarity with the surface of their target. The global mapping campaigns (Chapter 2) that made this possible also uncovered many of the asteroid's most scientifically exciting qualities. This chapter presents an abridged collection of maps illustrating Bennu's terrain, chemical make-up, temperature, and geology at the global scale.

At the foundation of those efforts was the global basemap — a mosaic of the entire asteroid constructed from more than two thousand PolyCam images — which serves as the authoritative visual reference of Bennu's surface. A map or flat picture of Bennu necessarily involves some distortion or omission, as with any two-dimensional representation (or projection) of a three-dimensional object. The global basemap is shown on the preceding pages, and below, in equirectangular projection, which causes less distortion than some once-widespread map projections, such as the Mercator. Nevertheless, it does stretch the polar regions, making them appear to have more area than they actually do and warping the shapes of features. Polar projections, though only showing one hemisphere at a time, preserve the shapes of features. This projection can be seen on the next two pages, as well as in close-up views of the basemap shown at the end of this chapter.

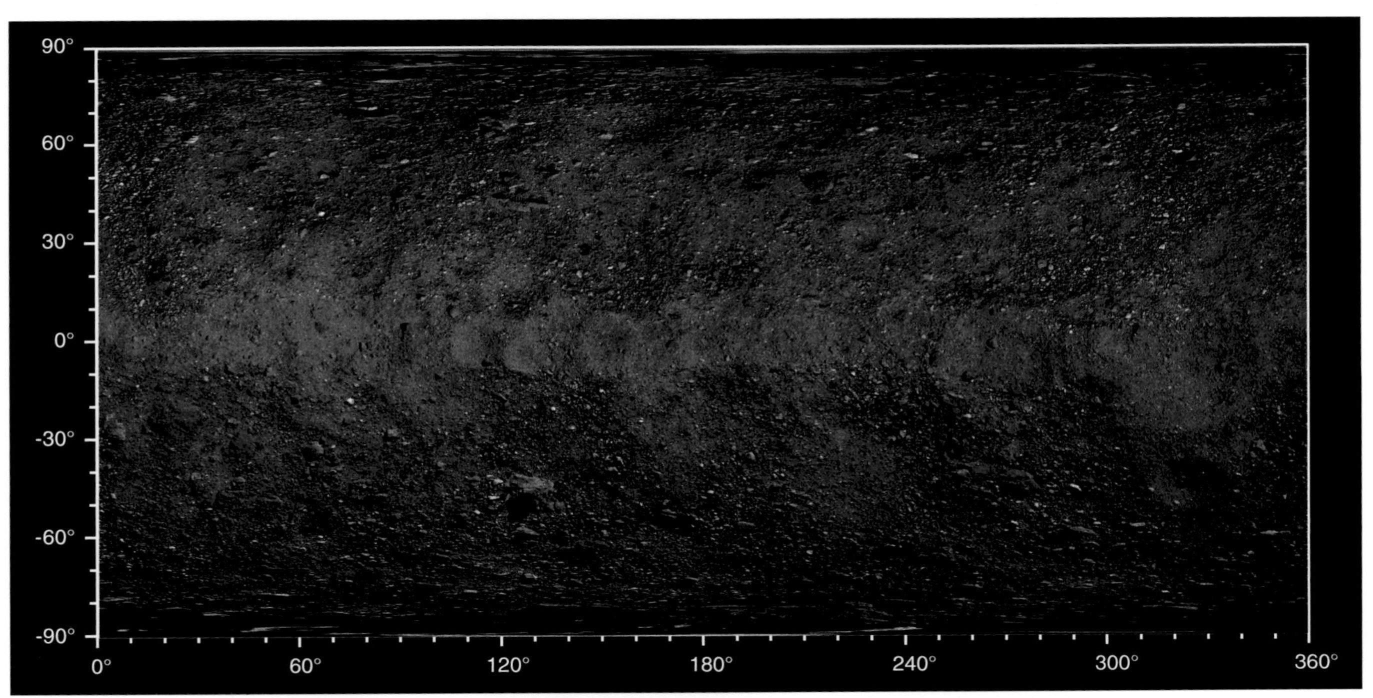

The global basemap of Bennu in equirectangular projection with coordinates.

Northern hemisphere basemap

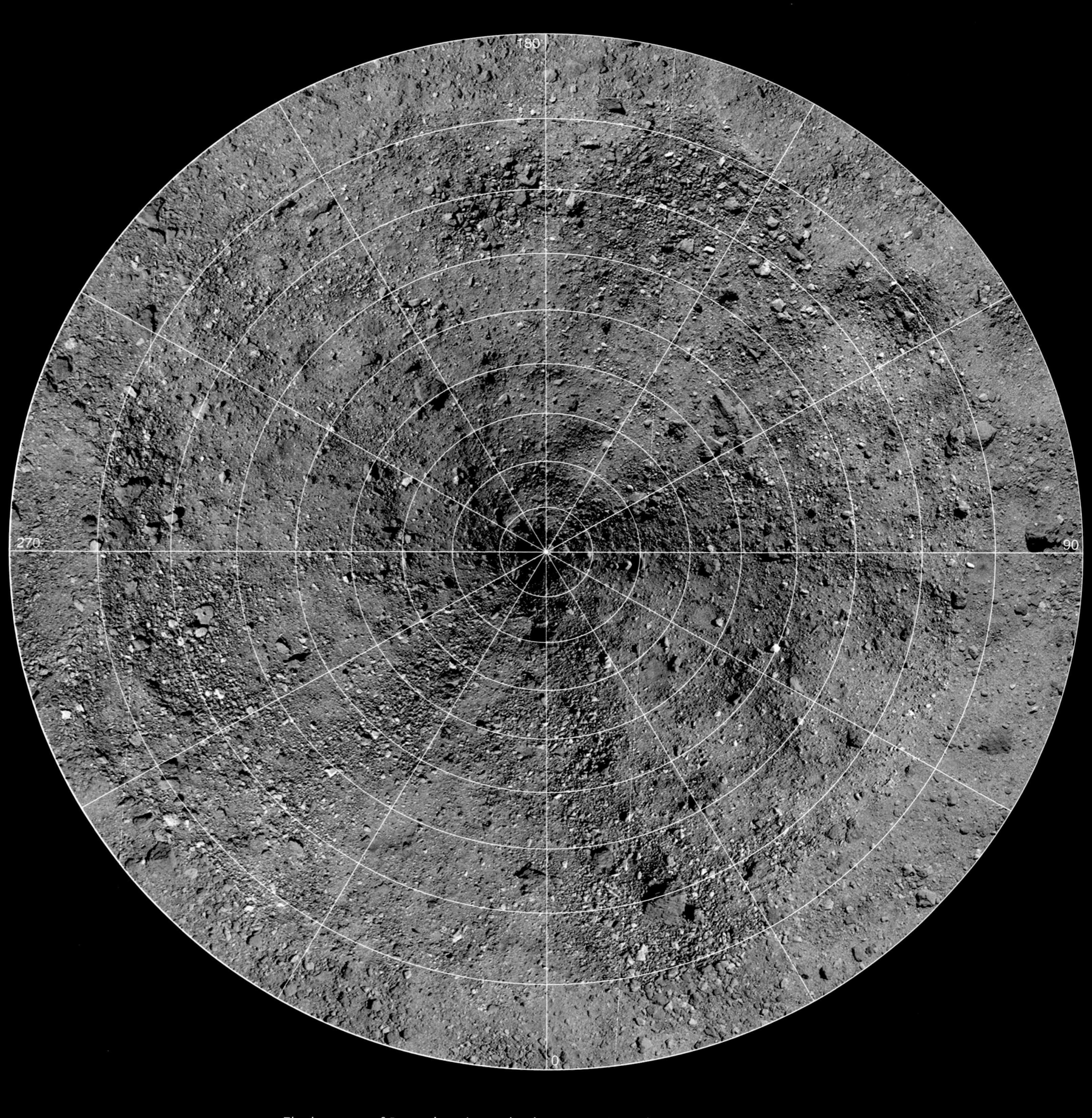

The basemap of Bennu in polar projection. For each hemisphere, the pole is at the center, and the equator is the outer edge. Lines of latitude (concentric, every 10°) and longitude (radial, every 30°) are overlaid.

Southern hemisphere basemap

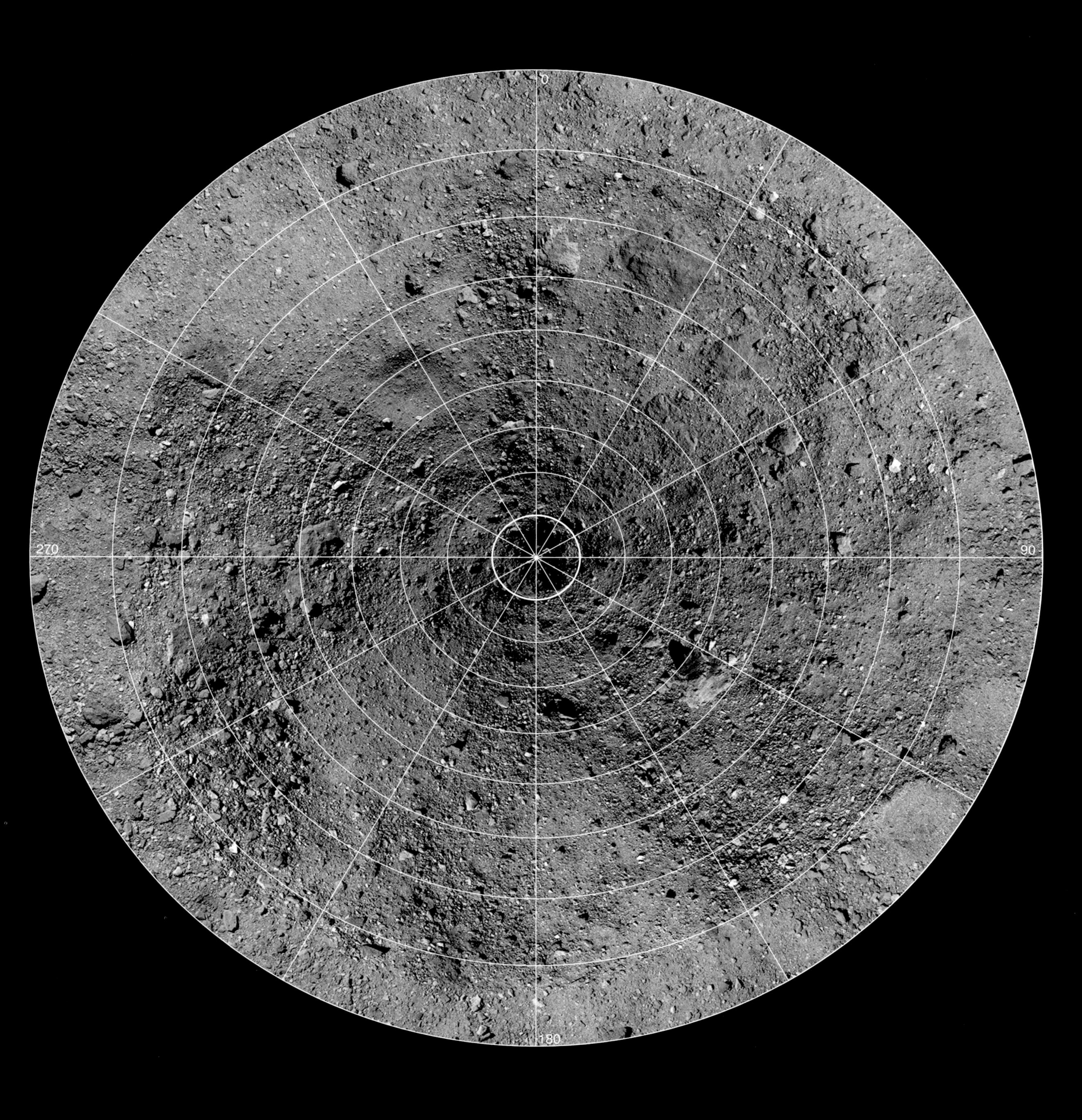

Coordinate system

"North" is defined differently for asteroids and comets than it is for planets. For a planet, the north pole is on the same side of the invariable plane (close to the plane of the ecliptic) as the north pole of Earth. On asteroids and comets, however, the axis of rotation, and thus the poles, can move from one side of the ecliptic to the other within decades. Therefore, for these bodies, the north pole is defined by the right-hand rule, illustrated below: If the right hand is in a fist with the thumb raised, and the fingers curl in the direction of rotation, then the right thumb is pointing north — that is, toward positive latitudes. Bennu has a retrograde rotation, meaning that unlike Earth, it spins in the direction opposite to its orbit and the rotation of the Sun.

Maps in this atlas use positive east longitude and planetocentric latitude. Positive east longitude is defined relative to the prime meridian, which the OSIRIS-REx mission placed at an easily recognizable boulder (Simurgh Saxum; Chapter 4). Planetocentric latitude is the angle between the equatorial plane and a vector from the center of mass of the body to a point on the surface. Latitudes in the northern hemisphere are positive, and those in the southern hemisphere are negative.

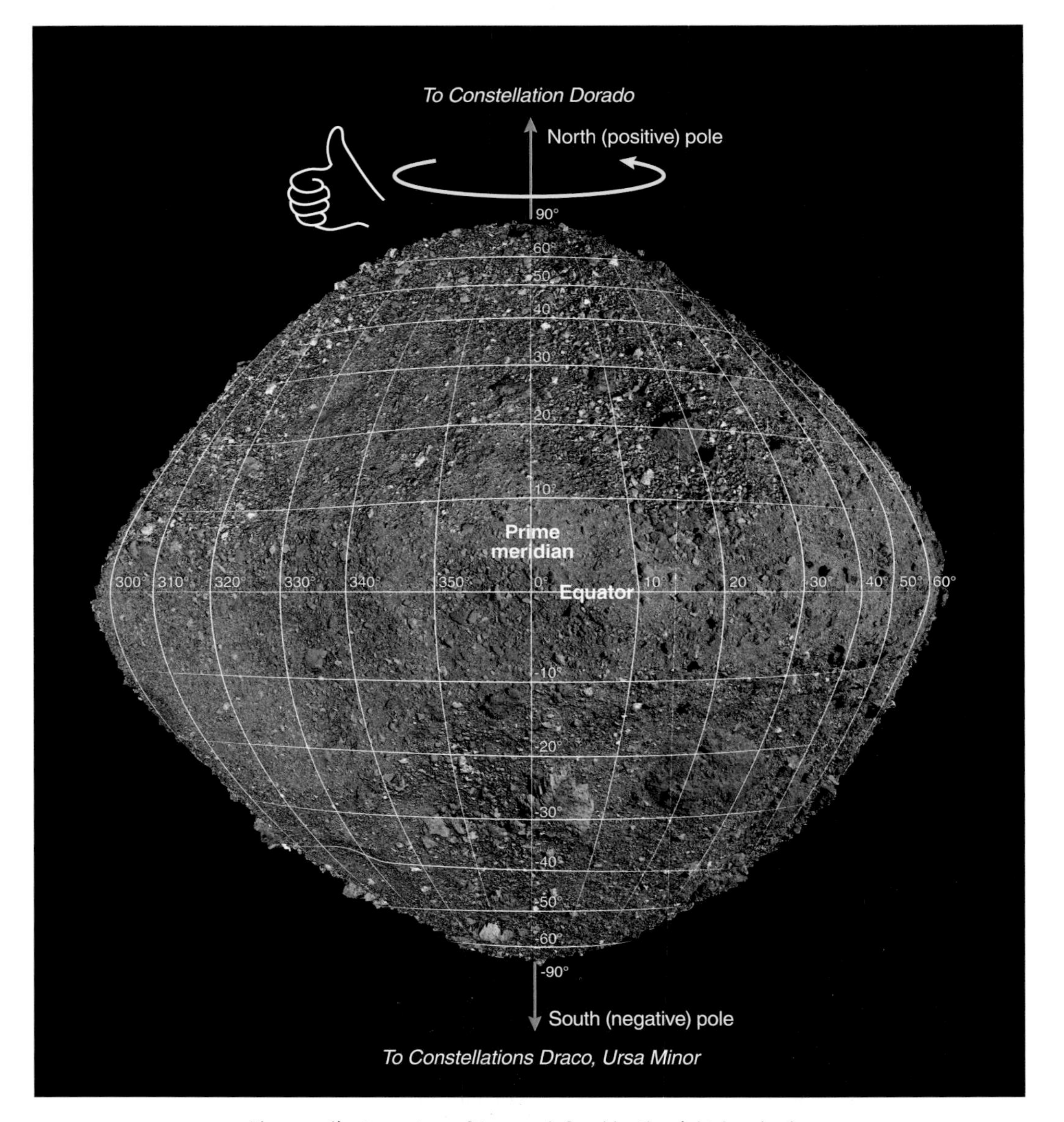

The coordinate system of Bennu, defined by the right-hand rule.

Effects of lighting

OSIRIS-REx mapped Bennu under a variety of lighting conditions, each of which offered its own advantages. For instance, in the images composing the basemap, the angle between the Sun, Bennu, and the camera (called phase angle) is uniformly about 30°. This angle was chosen to have enough shadow that surface features would be readily apparent, but not so much shadow as to obscure significant areas around larger features.

At higher phase angles, where the surface is mostly dark (as when viewing the crescent Moon), shadows dramatically enhance the circular crater rims and other high-relief terrain. At narrower phase angles, where the surface is mostly illuminated (as when viewing the full Moon), differences in the reflectivity of the surface materials can be perceived (next page).

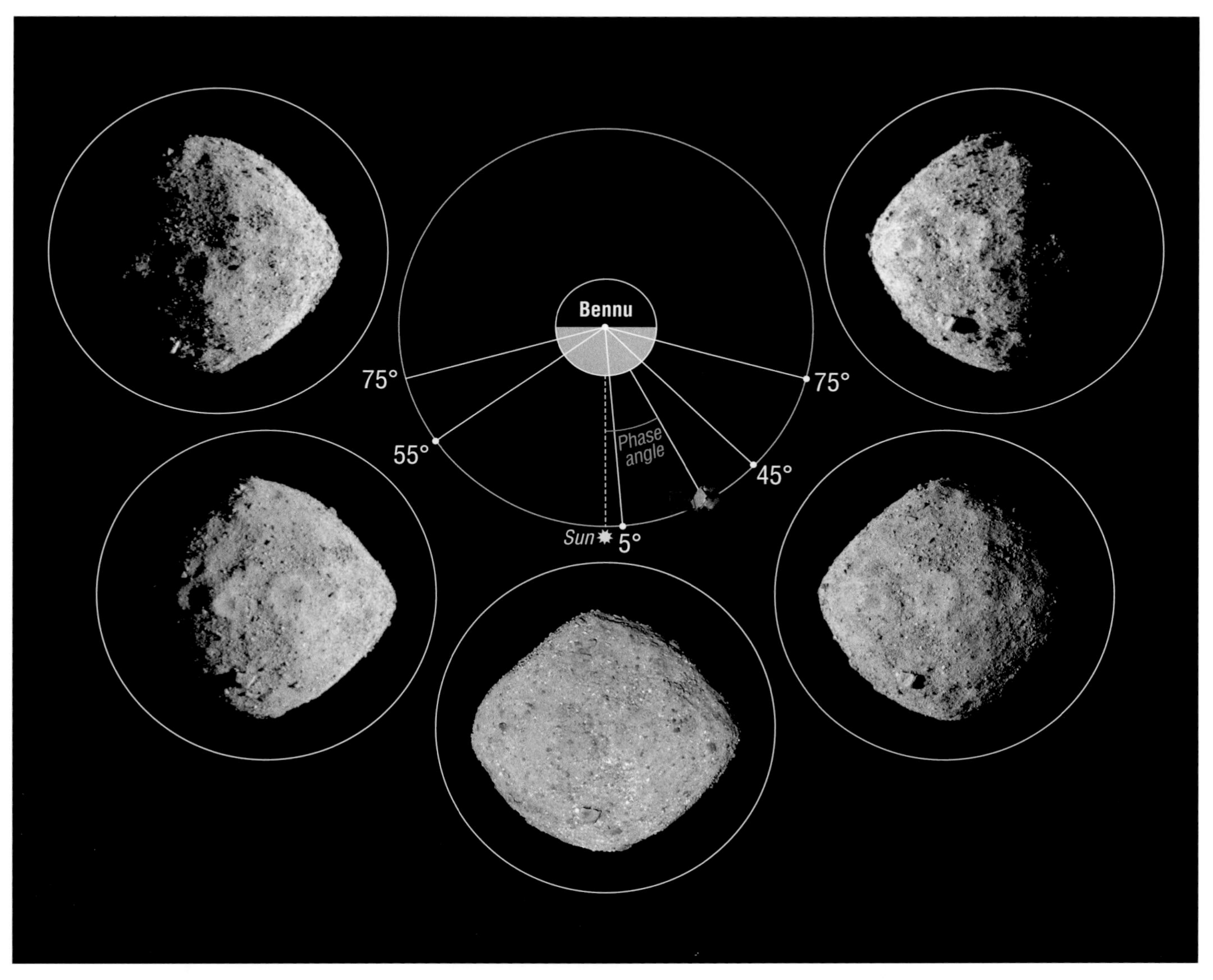

MapCam images of approximately the same face of Bennu under a range of phase angles. The shadow of BenBen Saxum, the prominent boulder in the southern hemisphere, lengthens at wider phase angles.

Reflectivity of the surface

Like the basemap, the map below is a mosaic of PolyCam images, except these were taken at a narrow phase angle of about 8° to minimize shadowing; note the disappearance of the crater rims. This map is therefore able to capture variations in the reflectivity, or albedo, of the surface materials. On average, Bennu is very dark, reflecting only about 4.5% of the sunlight that strikes it, whereas Earth reflects about 30%. Boulders tend to fall into one of two archetypes: either darker or brighter than the average "normal" albedo of the surface (for a zoomed example, see Odile and Odette Saxa, Chapter 4).

Reflectivity can also be assessed by using a spectrometer such as OVIRS to measure the proportion of sunlight reflected at a given wavelength (opposite page). But a limitation of PolyCam and OVIRS is that they cannot cover the shadowed polar regions. An alternative approach is to measure the intensity of laser pulses reflected back to OLA (Chapter 2) by Bennu's surface. Instead of relying on sunlight, OLA is an active sensor that creates its own photons, which it can fire uniformly over the global surface.

However, factors such as the slope, roughness, or porosity of the surface can also influence the reflected laser intensity. This might be the cause of the seemingly darker band along the equator in the OLA reflectivity map that is absent in the PolyCam map.

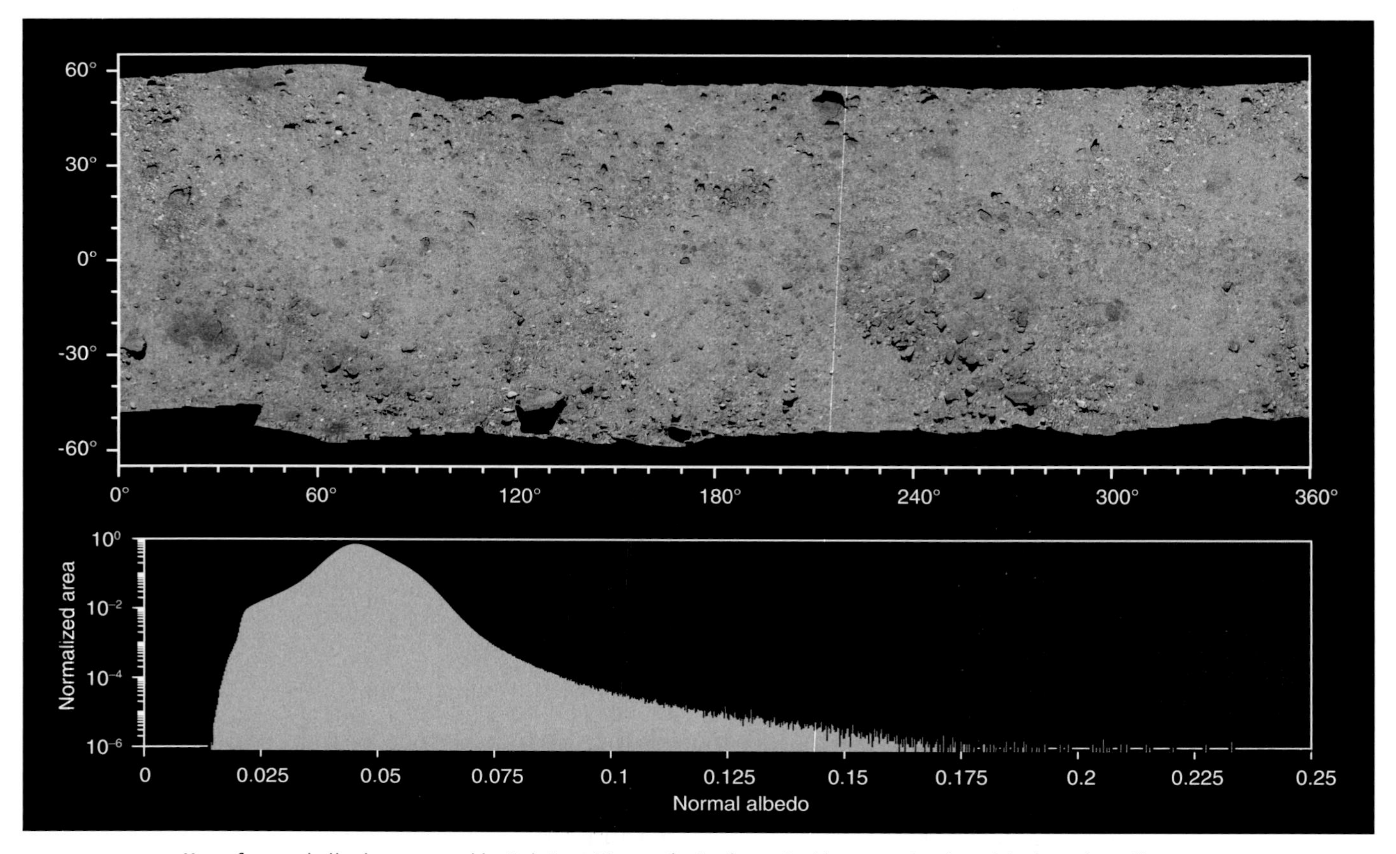

Map of normal albedo measured by PolyCam. The vertical axis on the histogram is a logarithmic scale to illustrate the full range of albedo values, especially the small population of extremely bright objects.

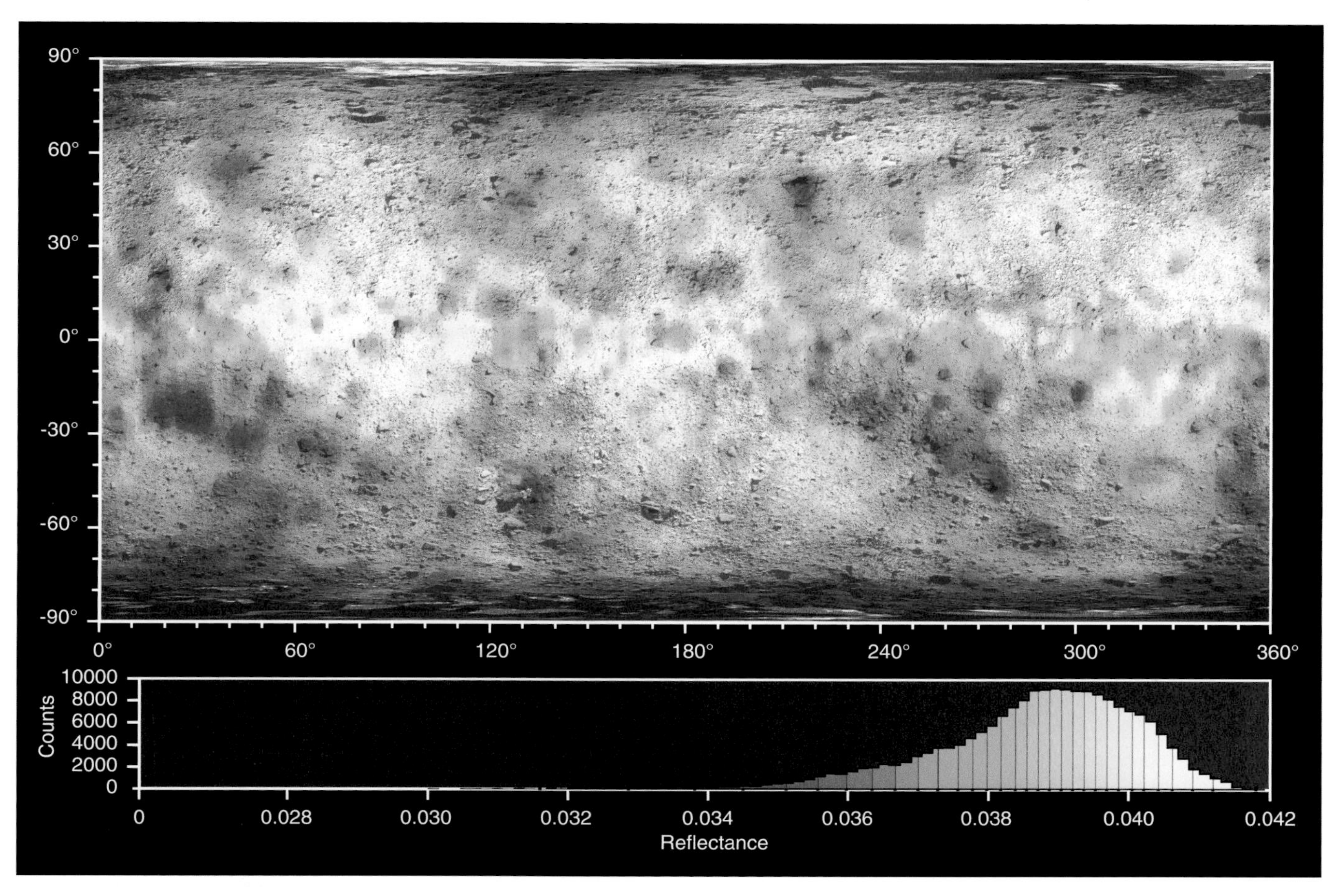

Map of reflectance at 0.55 µm, measured by OVIRS.

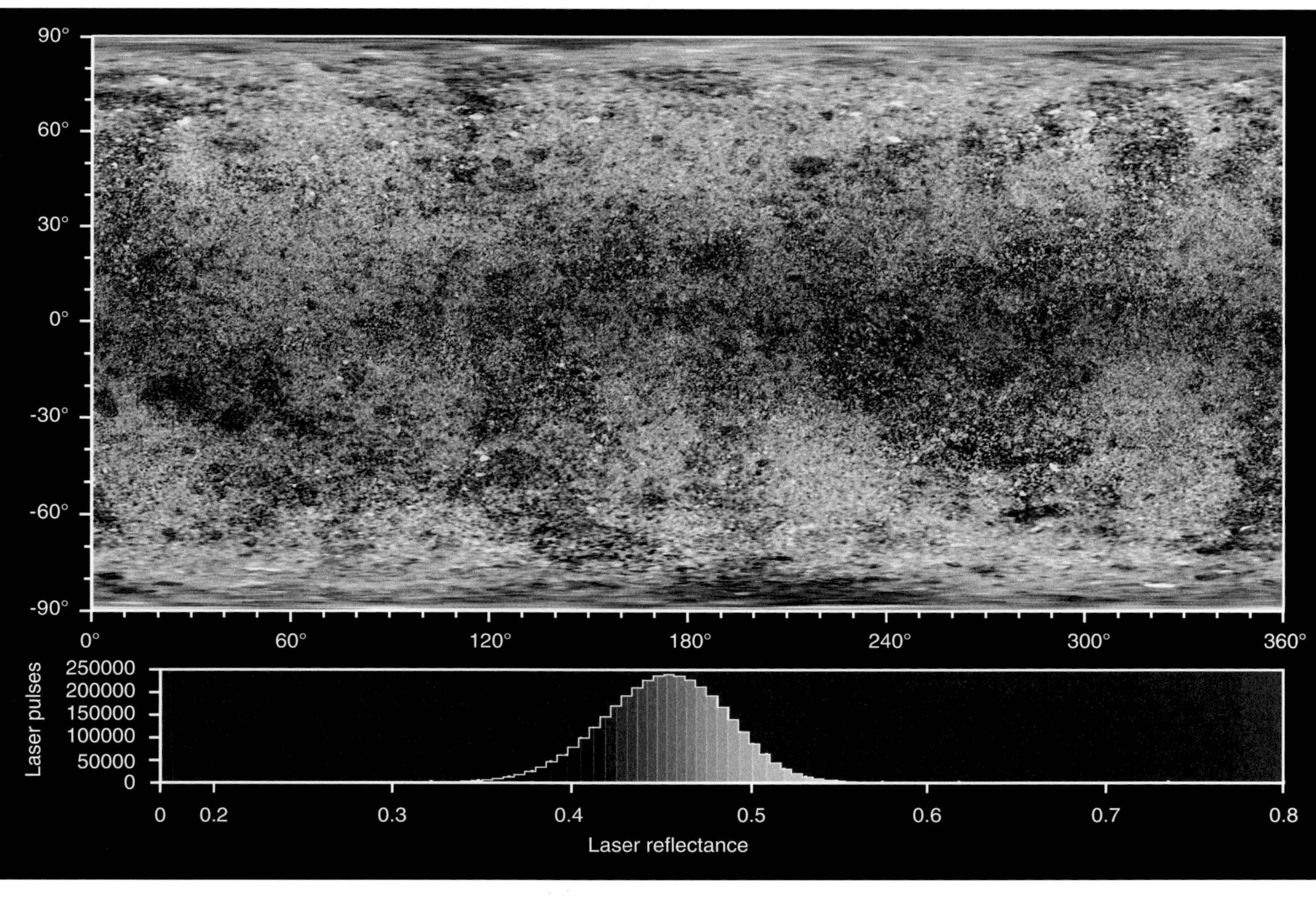

Map of reflected laser intensity measured by OLA. The laser photon's wavelength is 1.064 µm.

Bright fragments of other asteroids

Several unusually reflective boulders — five times brighter than the average surface — caught the team's attention during the initial global observations. They are difficult to discern in the global albedo map owing to their small size, so their locations are marked with letters on the basemap below, and they are shown in close up beneath the map.

MapCam and OVIRS observations of these areas indicated the presence of pyroxene, a mineral found in igneous rocks. This was a thrilling discovery because Bennu was never molten; therefore, the bright rocks must have originated elsewhere in the Solar System. Hundreds more of these potentially exogenic rocks and clasts have been detected (unlabeled white dots), though they too are so small that they constitute only a tiny fraction of Bennu's surface.

Based on the type of pyroxene detected, the most likely source of these foreign objects is the asteroid Vesta or its collisional descendants, known as the vestoids or V-type asteroids (Chapter 1). However, vestoid impactors large enough to leave behind the observed boulders would have destroyed Bennu in the process. More probably, the vestoids hit Bennu's larger predecessor (Chapter 3) and mixed with its surface material. After a later catastrophic impact, Bennu coalesced from the rubble and inherited the mixture.

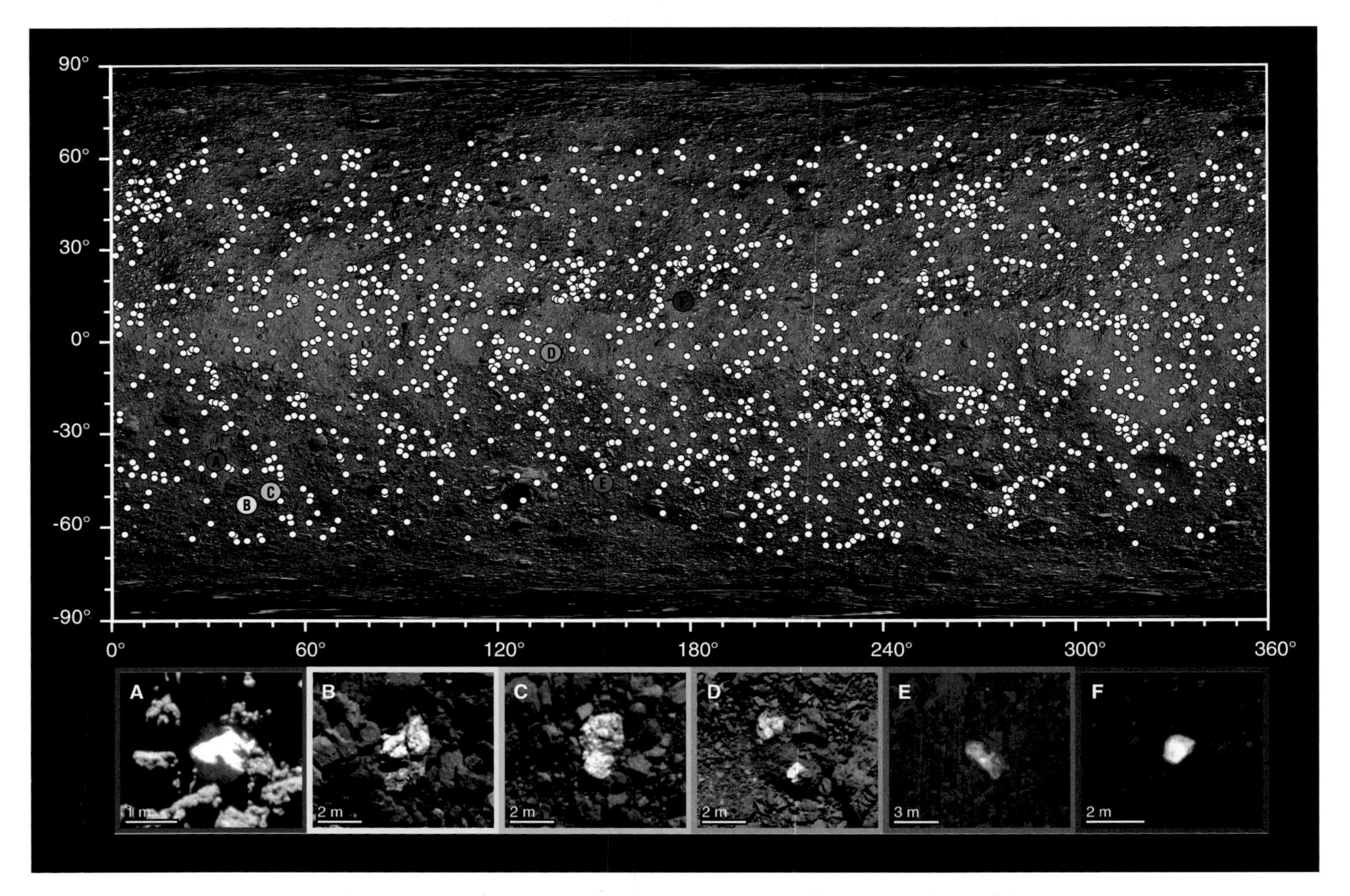

Map of anomalously bright rock fragments, represented by white dots, that likely originated from other asteroids. Images of six prominent examples are shown beneath.

Insights from spectral color

Isolating reflectivity in specific wavelength ranges reveals more about the surface than can be seen with the naked eye. In the false-color map on this page, red, green, and blue each correspond to the respective intensities of different spectral characteristics detected by MapCam (Chapter 2).

The takeaway from these spectral "colors" is that Bennu's surface materials have diverse geologic histories. On this map, dark boulders show up as red, and bright boulders as blueish, indicating a difference in their chemical and/or physical make-up. These two main rock types are inferred to originate from different environments on the parent asteroid.

Complicating matters, spectral color also varies with the degree to which a material has been weathered by the space environment. On Bennu there is no atmosphere to buffer the extreme solar radiation or the steady bombardment of meteoroids. The least weathered regolith, found in the basins of freshly formed craters (such as the Nightingale site), appears red in this map. Richer shades of blue — for example, the rock face at approximately −45° latitude, 265° longitude — also can represent fresh, unweathered material. In both cases, weathering over a long period of time changes the material to the predominant "shade" of medium blue that dominates the map.

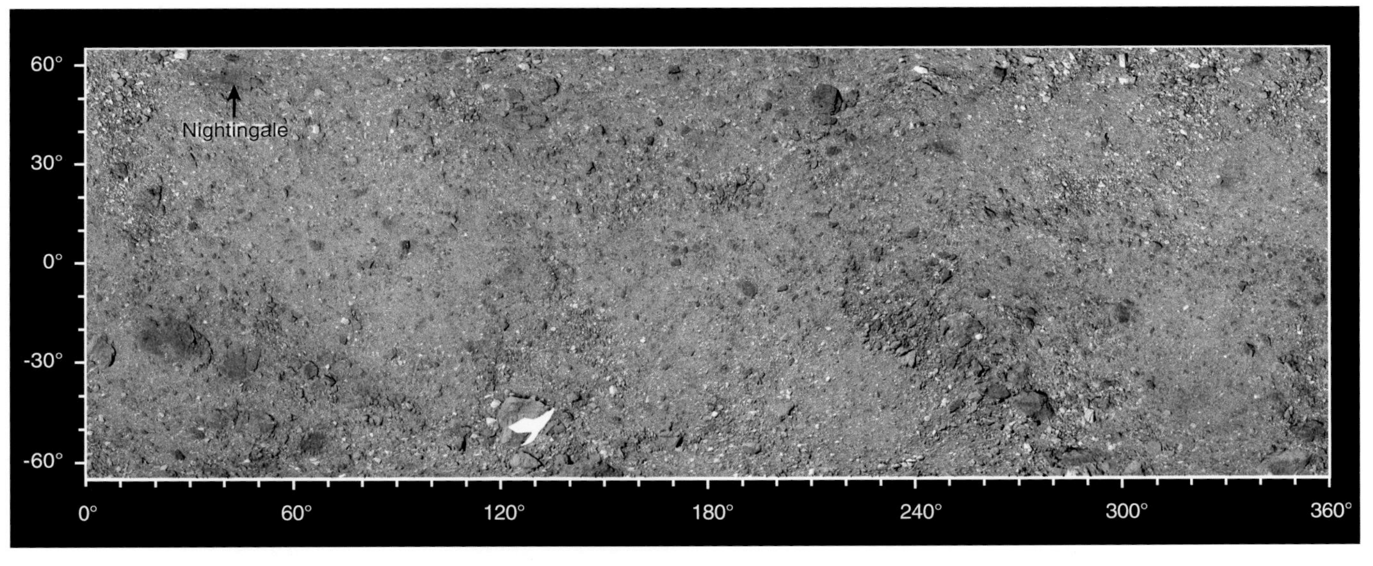

False-color map of spectral signatures measured by MapCam. Blue represents the steepness of the near-ultraviolet spectral slope; red, the steepness of the mid-visible to near-infrared slope; and green, the intensity of a spectral absorption at 0.7 μm that signals the presence of iron in the clay minerals. The white patch near the bottom of the map has no available data due to the massive shadow of BenBen Saxum.

Spectral detection of water

One of the OSIRIS-REx mission's first and most exciting discoveries was the existence of water on Bennu. Owing to the lack of atmosphere, no liquid water can exist. Rather, clay minerals from the early Solar System carry water in their crystal structures as hydroxyl ions. This finding lends credence to the hypothesis that long-ago collisions with primitive asteroids could have delivered the water that makes our planet habitable.

The hydroxyl ions strongly absorb infrared light at wavelengths around 2.7 micrometers, creating a spectral feature known as the hydration band. OVIRS detected a strong hydration band as soon as Bennu filled its field of view. Later, global surveys revealed how hydration varies regionally, as illustrated below. The map shows that the entire surface is hydrated, but the equator, where Sun exposure and peak temperatures are highest, is drier than elsewhere.

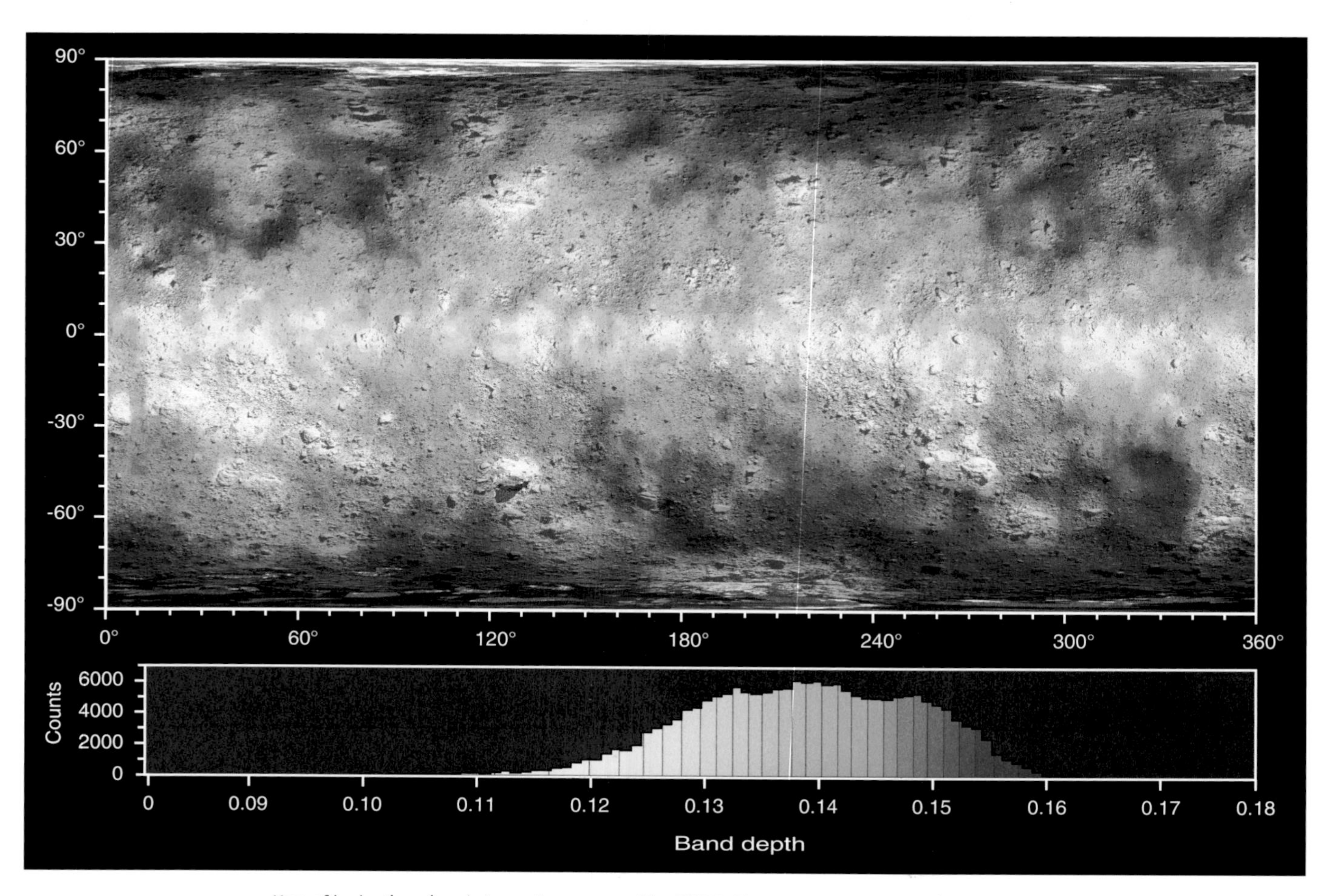

Map of hydration signal strength, measured by OVIRS. The parameter mapped is the depth of the hydration band in the spectrum (Chapter 3) as a proportion of the total reflected sunlight. Cooler colors (darker blues) indicate higher water abundances.

The building blocks of life

OVIRS also detected widespread carbon-bearing compounds, including the organic molecules upon which all known life is based. The existence of such molecules implies that primitive asteroids could have seeded the early Earth not only with water, but also with the relevant ingredients for the origin of life.

Other carbon-bearing compounds on Bennu include carbonate minerals such as calcite and dolomite that make up limestones and evaporation crusts on Earth. Although these are not directly pertinent to life, they indicate that heated liquid water circulated through Bennu's parent asteroid (Chapter 3). Such conditions are analogous to hydrothermal vents, which could have altered or created new organic compounds essential to life on Earth.

Organic molecules and carbonate minerals share a spectral absorption feature around 3.4 micrometers, sometimes called the carbon band. The strongest carbon signal tends to come from individual boulders, including those surrounding the Nightingale site (Chapter 5). This finding gave the team confidence that the mission's goal of collecting a carbon-bearing sample would be achieved.

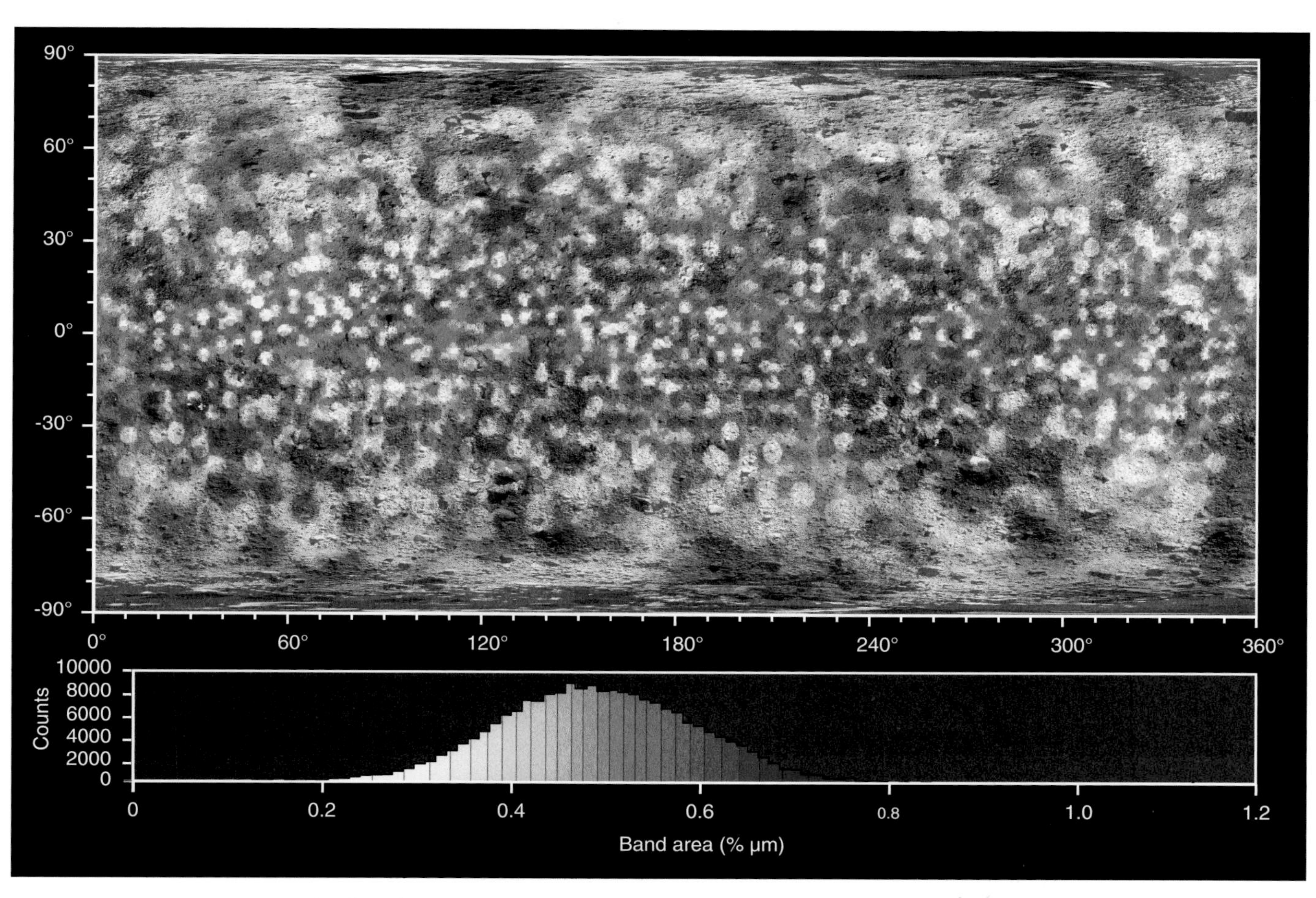

Map of carbon signal strength, measured by OVIRS. The parameter mapped is the area of the carbon band in the spectrum, as a percentage of the total reflected sunlight. The spotted appearance is an artifact of the circular area OVIRS "sees" with each measurement.

Sizes and ages of craters

The pattern of craters on a planetary surface records its history of collisions with other objects. Scientists have a good idea of the size and frequency of projectiles that Bennu would have encountered during both its time in the main belt and its current near-Earth orbit. This information, combined with knowledge of how the surface responds to impact (Chapter 6), makes it possible to estimate how long ago the craters formed.

To map the numerous craters on Bennu, the OSIRIS-REx team partnered with citizen scientists. Ultimately, more than 1,500 craters were cataloged. Bennu would probably have even more craters if not for its many boulders, which act as armor against small impacts.

Each ellipse on the basemap below outlines a mapped crater, color-coded by size. The apparent elongation of craters near the poles is an artifact of the map's inherent distortion.

Estimated crater ages for Bennu range from about a billion years for the largest to less than 100,000 years for the smallest (red in the spectral color map). This extreme diversity of ages suggests a dynamic, constantly evolving landscape, parts of which have been resurfaced multiple times by geologic processes.

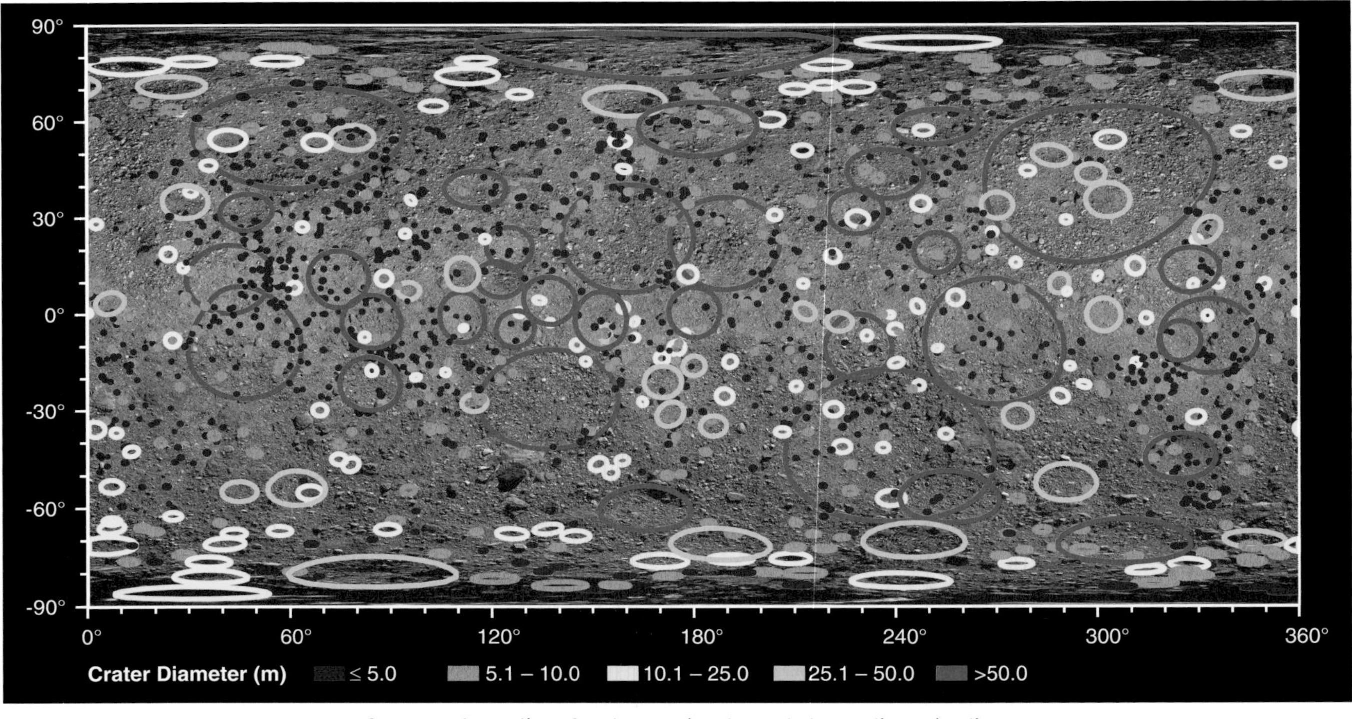

Map of craters. The outline of each crater is color-coded according to its diameter to visualize how craters of different sizes are distributed geographically.

Surface temperature and resistance to heat

To probe the temperature on the surface of Bennu, the OSIRIS-REx mission relied primarily on thermal emissions sensed by OVIRS and OTES (Chapter 2). These measurements were first fed into a computational model to calculate the thermal inertia of the surface materials. Thermal inertia, mapped below, describes how resistant a substance is to changes in temperature. On Earth, solid rocks have high thermal inertia, meaning they tend to warm up and cool slowly. By contrast, sand has low thermal inertia; it heats up quickly in the Sun and cools down quickly in shadow.

Bennu's thermal inertia is low overall, but it varies geographically, unexpectedly reaching its lowest on large, dark boulders, such as Roc Saxum. From this apparent reversal of the norm, the team inferred that these boulders must be extraordinarily porous and weak compared with rocks on Earth. Brighter boulders typically have somewhat higher thermal inertia, so they may be at least marginally denser and stronger. The highest thermal inertias are concentrated along the equator, possibly due to higher-strength material surviving transport to this region.

Thermal inertia is the key to predicting surface temperatures. With this property mapped, the team could move on to calculating temperatures for any local time of day or year.

On the next page the maximum and minimum surface temperatures are mapped for the warmest time of year, which occurs at perihelion when Bennu comes closest to the Sun.

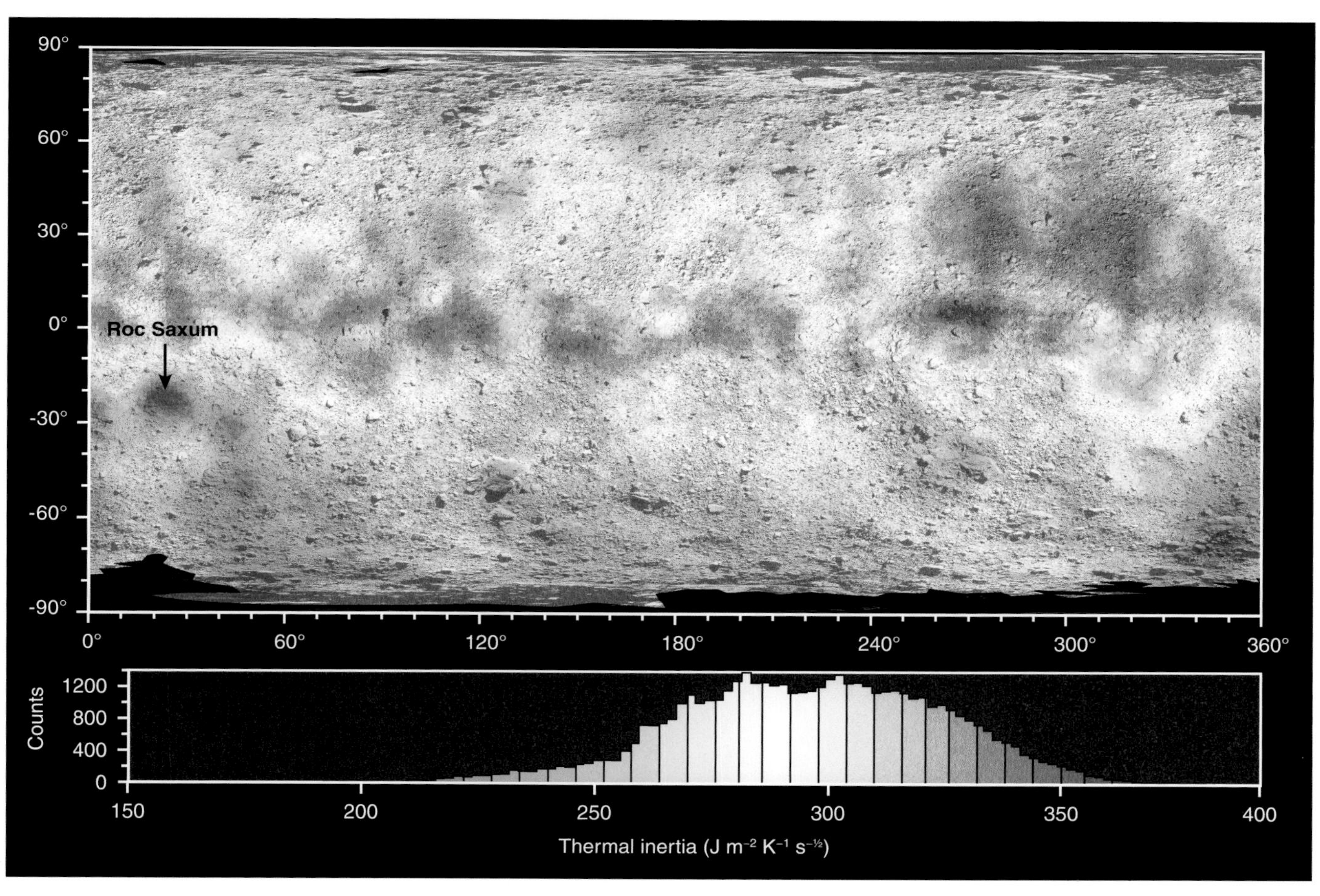

Map of thermal inertia, based on OTES measurements. The units of Joules per square meter per degree Kelvin per second square root are the international standard for measuring this property.

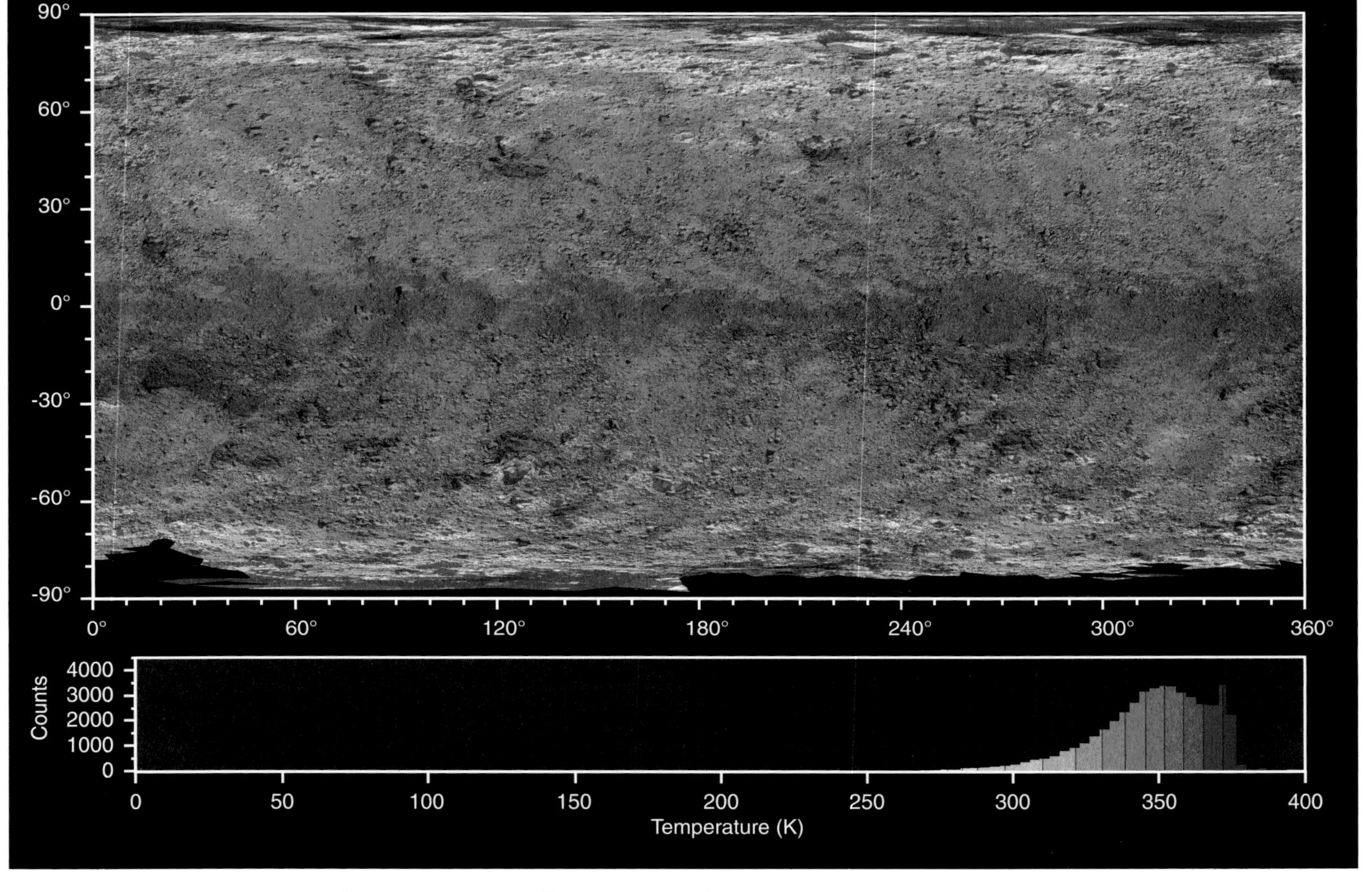

Daily high temperatures (in degrees Kelvin) at the warmest time of Bennu's year.

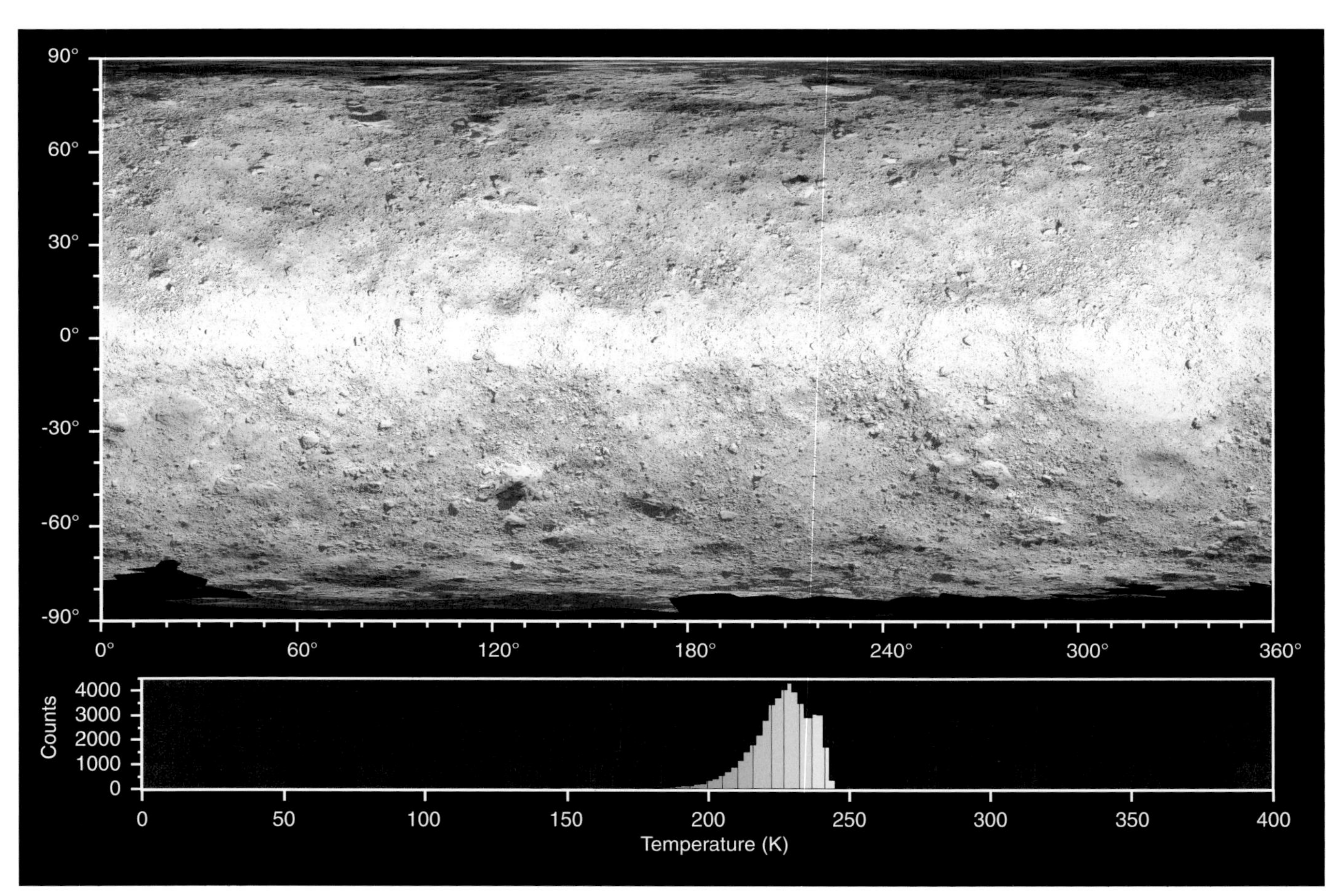

Daily low temperatures at the warmest time of Bennu's year.

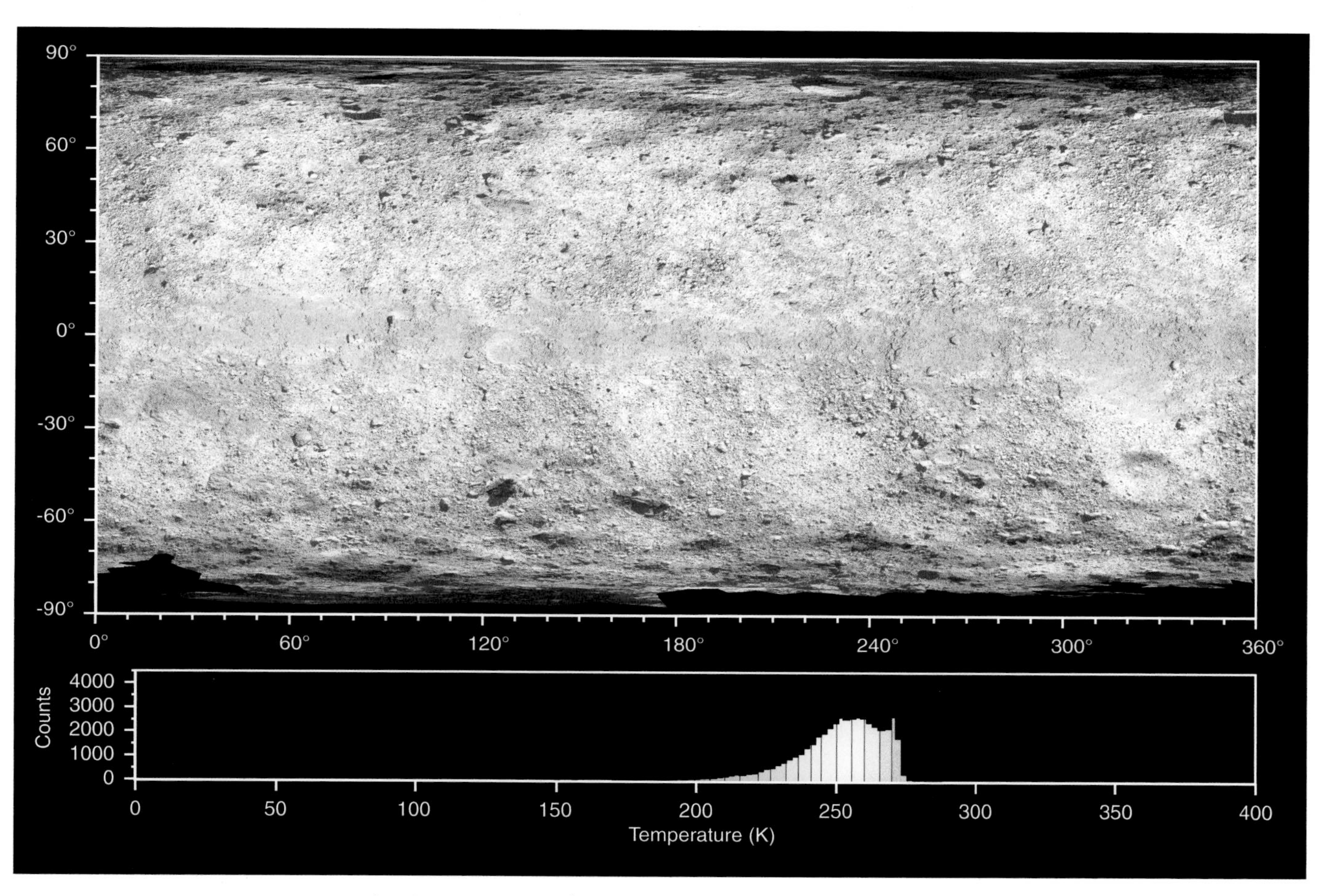

Daily high temperatures (in degrees Kelvin) at the coldest time of year.

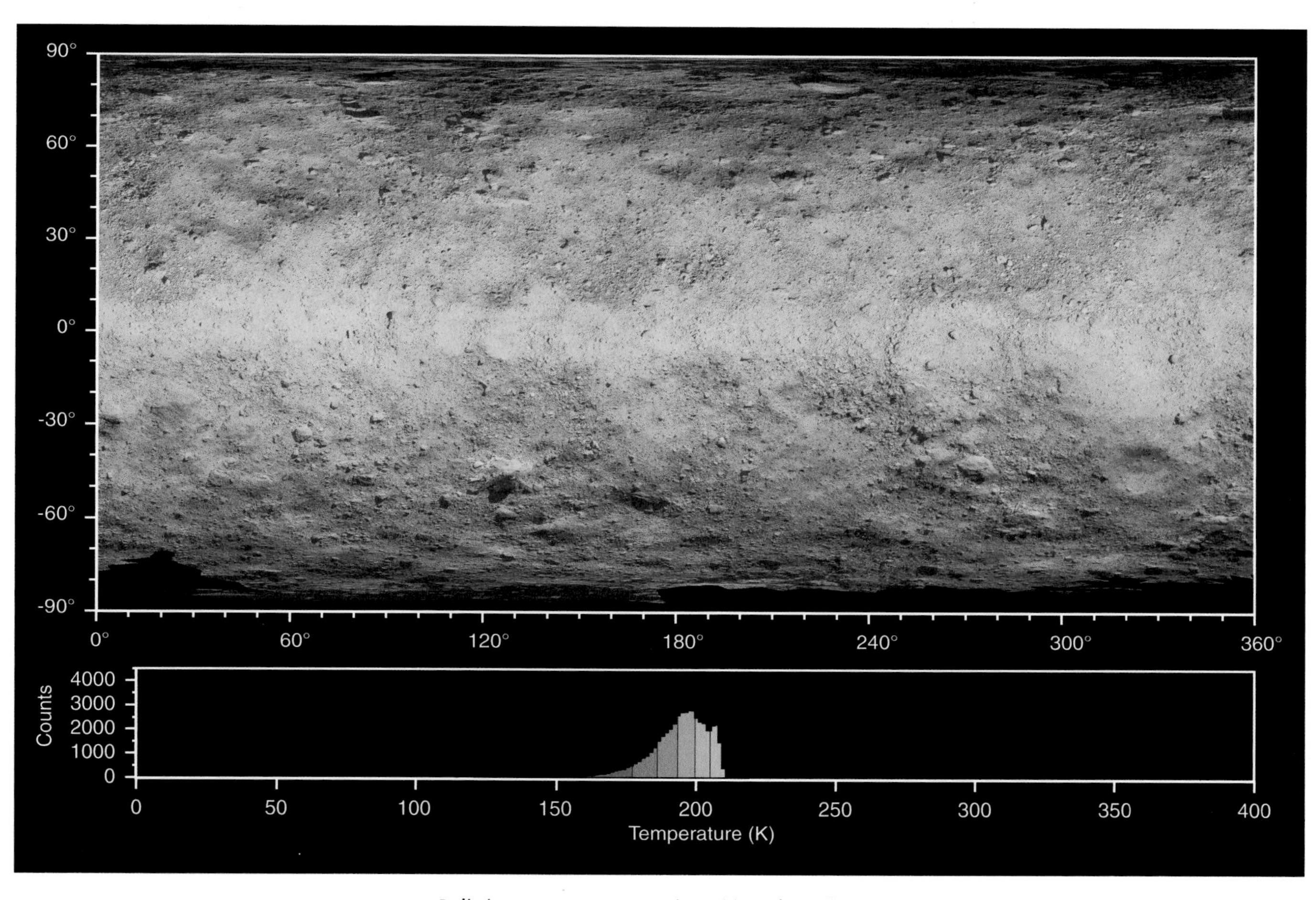

Daily low temperatures at the coldest time of year.

On the previous page the maximum and minimum surface temperatures are mapped for the coldest time, at aphelion, when Bennu is farthest from the Sun. Comparing the maps illustrates the dramatic seasonal changes in temperature that the surface undergoes over the 437-day orbit. Together, these maps also show that Bennu's poles stay relatively cold. By contrast, at the equator, the temperature swing is extreme: from hotter than boiling water by day to Antarctic cold by night.

Radius

One way to understand the shape and structure of an asteroid is to map the radius, or distance, from its center to each point on its surface. The radius map below confirms that Bennu's characteristic bulging equator is the widest part of the asteroid (red and orange) and shows that the narrowest regions are in fact between the equator and poles (cool colors), whereas the poles widen slightly.

Unlike the other maps in this chapter, the data here are shown by themselves, not as an overlay on the basemap. Yet some familiar features on the surface still appear, because they are large enough to create local increases in Bennu's radius. For instance, the red semicircular feature on the equator at about 270° longitude is the rim of Minokawa Crater. BenBen Saxum pops out in yellow at coordinates of about –45°, 125°.

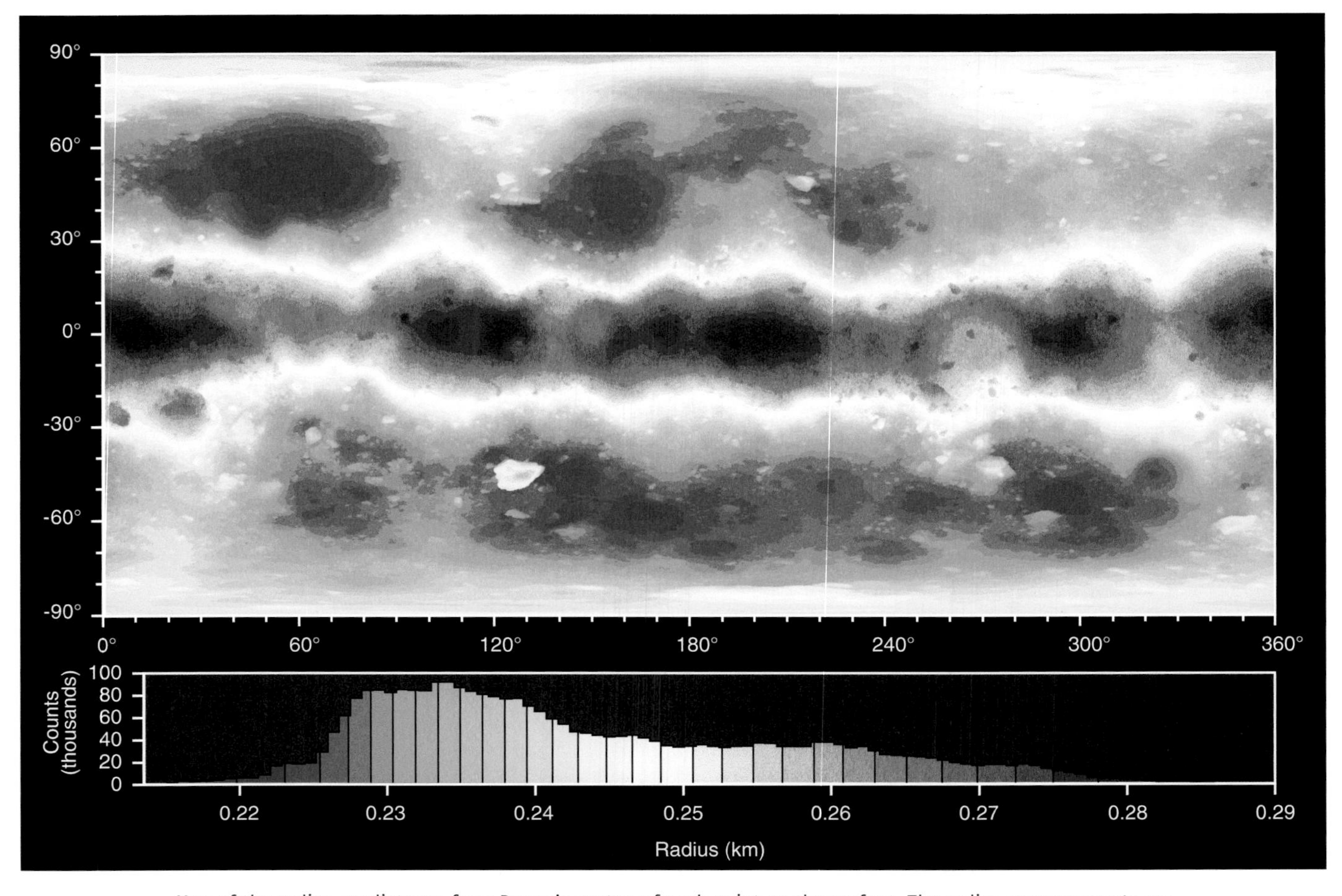

Map of the radius, or distance from Bennu's center, of each point on the surface. The radius measurements are drawn from a digital terrain model constructed using OCAMS images and OLA ranging data.

Geopotential elevation

Geopotential is a measure of the acceleration at the surface of a planetary body. Geopotential elevation, mapped here, indicates the combined influence of gravity and centrifugal forces across the surface. On Earth, centrifugal forces are weak compared with gravity, meaning that height and geopotential elevation are equivalent.

However, on Bennu, centrifugal forces are nearly as strong as gravity. The highest geopotential values are at the poles, and the lowest are at the equator. Because centrifugal forces at low latitudes counter the weak pull of gravity, Bennu effectively slopes downward from the poles to the equator.

As a result, regolith tends to migrate toward the equator in small landslide-like events. This gradual buildup of material may bury boulders, leading to the relatively smooth appearance of the equator in the basemap.

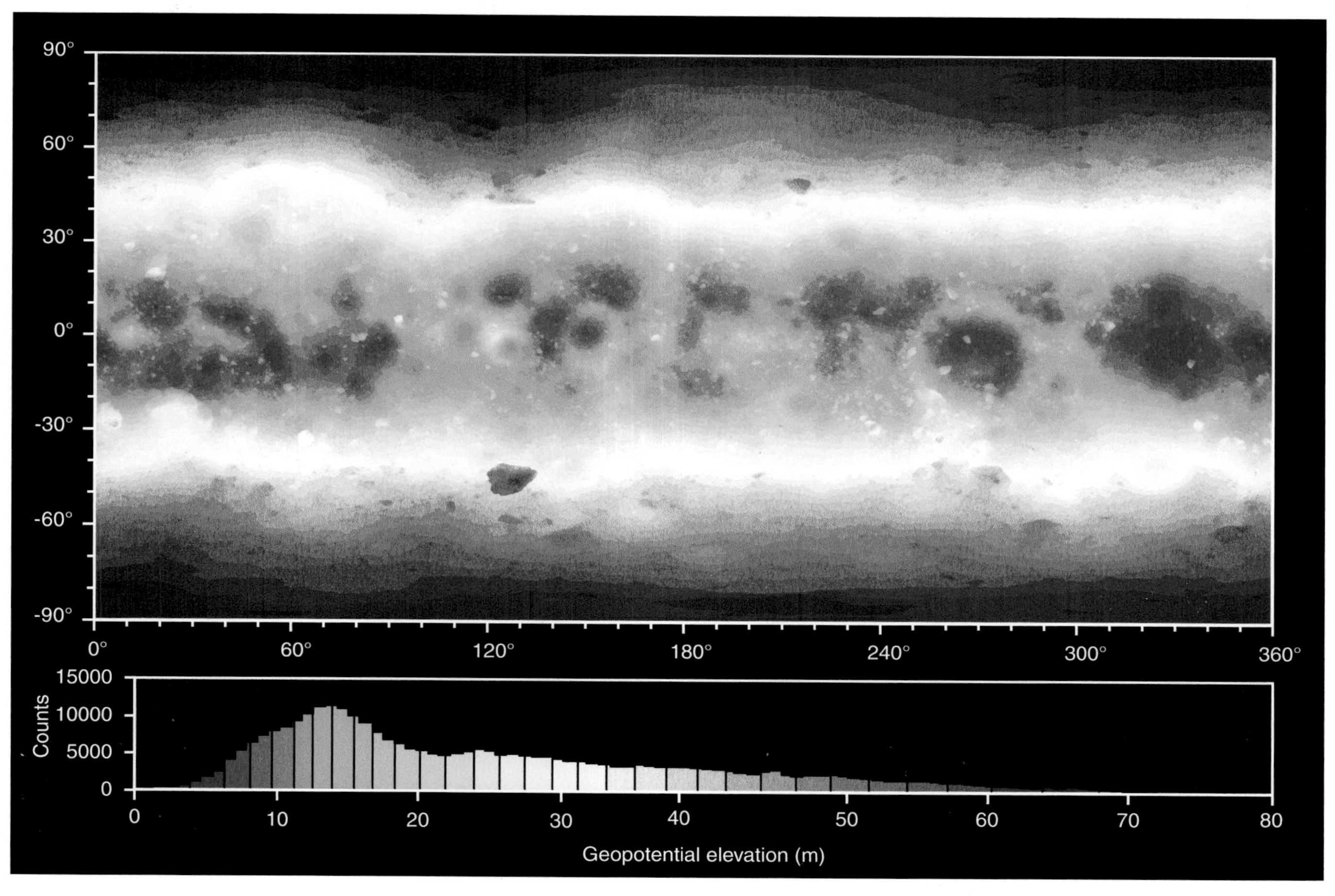

Map of geopotential elevation. As in the radius map, the data are shown without an underlying basemap; boulders and crater rims can be discerned as local increases in geopotential elevation.

Roughness of the terrain

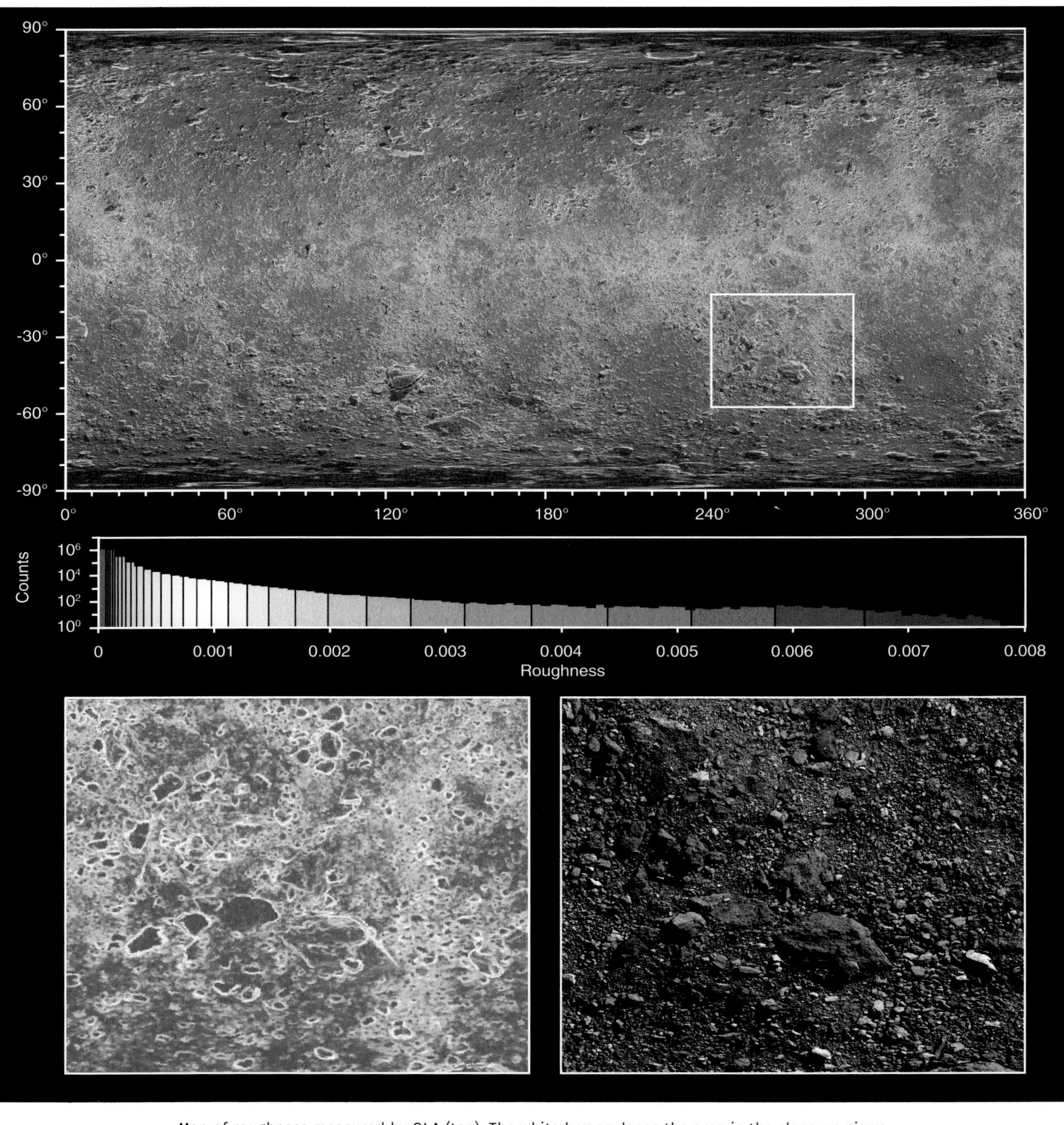

Map of roughness measured by OLA (top). The white box encloses the area in the close-up views below. Below left illustrates how accurately the roughness data conform to the edges of rocks. Below right is a close-up of the same area in the basemap, for comparison.

The high-definition laser ranging data acquired by OLA made it possible to produce digital models of the global terrain that were accurate to within 20 centimeters. This approach used reflected laser travel time, together with the known speed of light, to calculate the distance from the spacecraft to the surface for every laser pulse fired. OLA collected these ranging measurements every five centimeters; to cover the full surface, this meant firing the laser at Bennu about three billion times.

The resultant wealth of data offered a way to quantify and map the irregularity, or roughness, of the terrain. Roughness is calculated as the standard deviation in surface height in a 30-centimeter-squared area. The maximum roughness values hug the edges of steep rock faces; note the red border around Bennu's tallest boulder, BenBen Saxum (Chapter 4). Less extreme high roughness, in green and yellow, occurs in areas where dense concentrations of boulders create rugged terrain. The smoothest areas, in dark blue, contain the smallest boulders, and show low roughness values.

Bennu's surface can be described by two geologic units, intuitively called the Smooth Unit and the Rugged Unit. As the names suggest, the two units are primarily distinguished by the roughness of their terrain. They also differ in their respective crater populations, indicating different surface ages. The Smooth Unit likely dates to Bennu's time in the main belt, whereas the Rugged Unit may represent resurfacing after Bennu arrived in near-Earth space.

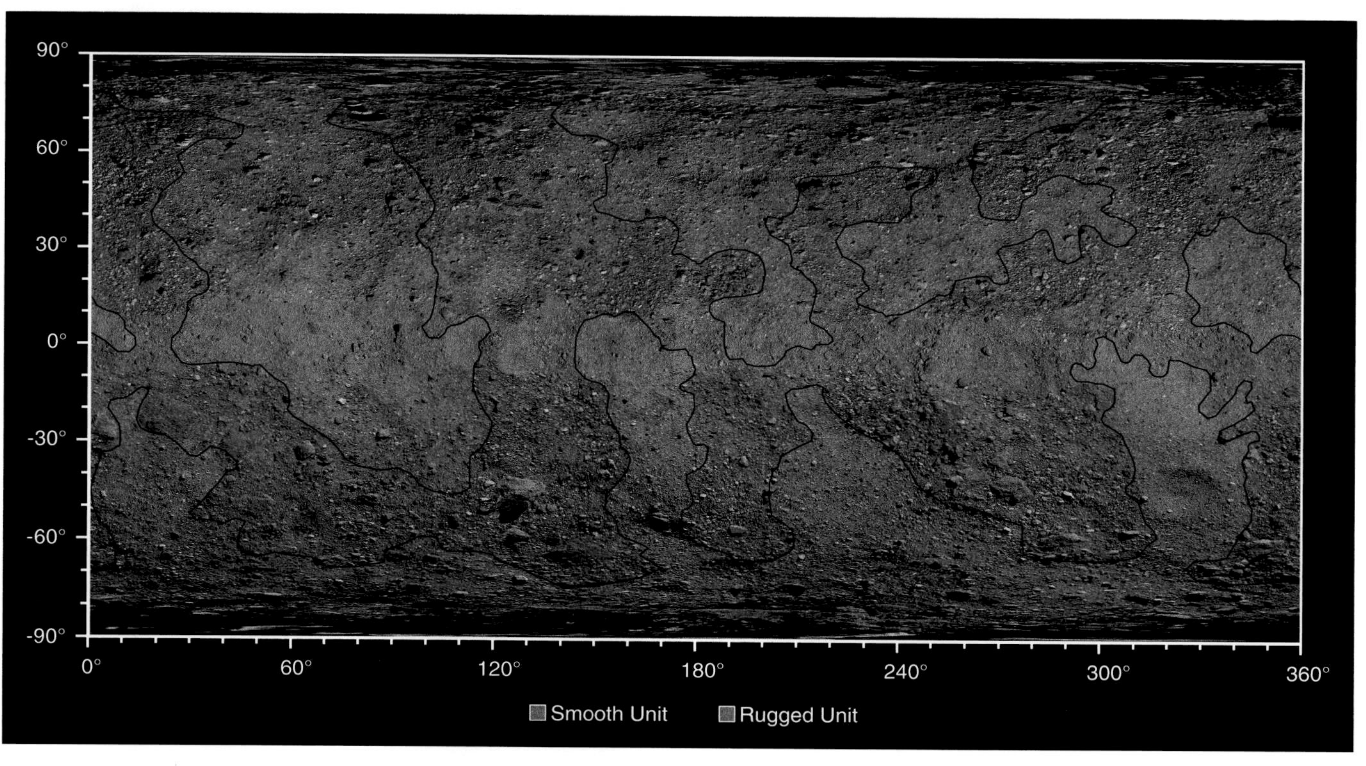

Geologic map delineating the older, smoother surface from the younger, more rugged one.

Gravity field and density

The gravity of a planetary body scales with its mass, which makes it challenging to measure for a small asteroid like Bennu. One approach is to detect subtle deviations in the spacecraft's orbit due to the pull of the asteroid's gravity. However, the weaker the gravity, the closer the spacecraft has to come for this technique to work. After arriving in the microgravity environment of Bennu, the team realized that OSIRIS-REx would have to draw dangerously close to measure the gravity field at the desired precision.

Fortunately, Bennu produces its own gravity probes, in the form of rock particles that are ejected periodically from the surface and sometimes enter a short-lived orbit (Chapter 3). By tracking these particles across navigation images, the team deduced Bennu's gravity field from their motions.

This derived map of gravitational anomalies shows how the measured gravity field differs from what would be expected if Bennu had a spatially uniform density. The geographic variation demonstrates that Bennu's mass is unevenly distributed throughout its volume — not surprising for an asteroid consisting of irregularly sized rubble and voids.

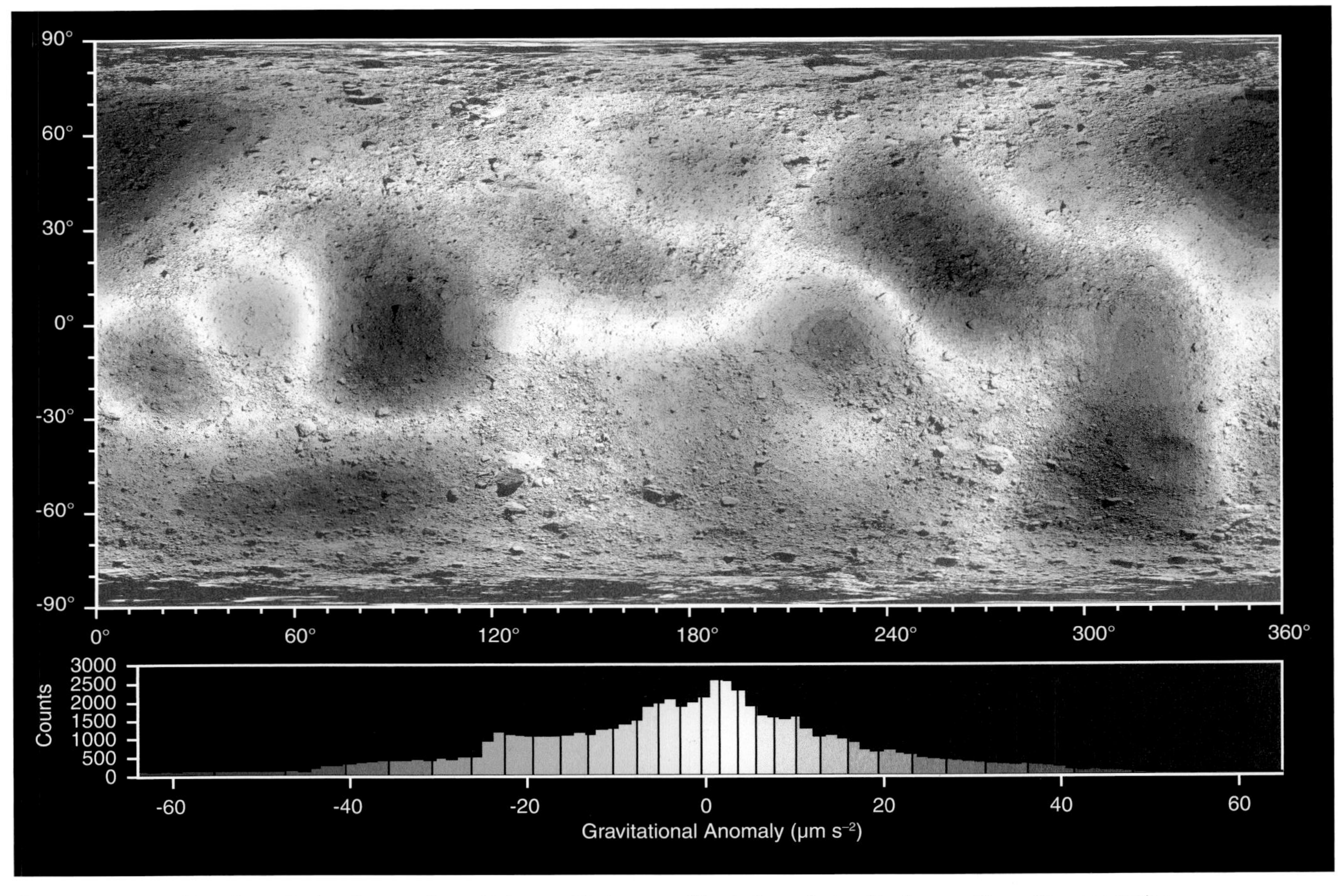

Map of gravitational anomalies, which reveal heterogeneities in Bennu's density. Areas with positive anomalies (warm colors) are denser than expected; those with negative anomalies (cool colors) are less dense than expected.

The global basemap – a closer look

Mapping a small, irregularly shaped body like Bennu presents unique challenges. How did the scientists and engineers on the mission construct the high-resolution, seamless image map of the entire surface? The 2,000-plus images composing the mosaic map were captured by the telescopic PolyCam imager during Detailed Survey Flybys 3 and 4 (Chapter 2). For each flyby, the spacecraft moved slowly while Bennu rotated and PolyCam nodded up and down, snapping pictures in a serpentine pattern across the surface.

With these images on the ground, the team first set about "controlling" them to remove uncertainty in the location of the pixels relative to Bennu's surface. Although we can estimate what a pixel's coordinates should be by knowing the positions of the spacecraft, camera, and asteroid, imperfections are inevitable with pixels five centimeters across imaged from kilometers away. If we were to lay two images next to each other using the initial estimation of their locations, there would be some misregistration, meaning the same rock might appear twice, next to or overlapping itself. When controlling the images, we can manually match that rock in two images and "pin" those pixels together, or manually match its relief in the 3-D shape model and pin its location as ground truth. With enough of these pinned points, we can apply best-fit calculations to the rest of the pixels across all the images and significantly reduce the uncertainty. This labor-intensive process produced a highly accurate global basemap.

Once control was established, the images were orthorectified, a process that removes geometric distortion, so that every pixel and feature looks as though viewed from directly above. Next, the team worked on making images taken on three different days look as uniform as possible. Because of Bennu's curves, the poles are darker and have longer shadows than the equator at a given local time. By understanding and modeling Bennu's light-reflecting behavior (page 151–152), we can adjust the brightness of the pixels in an image to match the amount of light they would reflect at a different time of day. This "photometric correction" makes every feature look like it was photographed with the Sun in essentially the same position. However, it does not change shadows, so they remain longer near the poles.

For a spherical body, the process of "mosaicking", or stitching together images to form a map, is straightforward. Because of Bennu's irregular shape, however, images from the three flybys had to be mosaicked separately, then carefully hand-trimmed on a feature-by-feature basis. This involved choosing which flyby represented the best view of a feature. To create the seamless finish, rather than cutting images along lines of latitude, the team hid the seams of the mosaic in shadows and natural feature boundaries, such as crater edges.

In the following pages, the basemap of Bennu is divided into 28 sections for detailed viewing. The location of each section is indicated in 2-D space on a thumbnail of the full basemap, and in 3-D space on a rotating view of the shape model with its surface colored by radius (as on page 162).

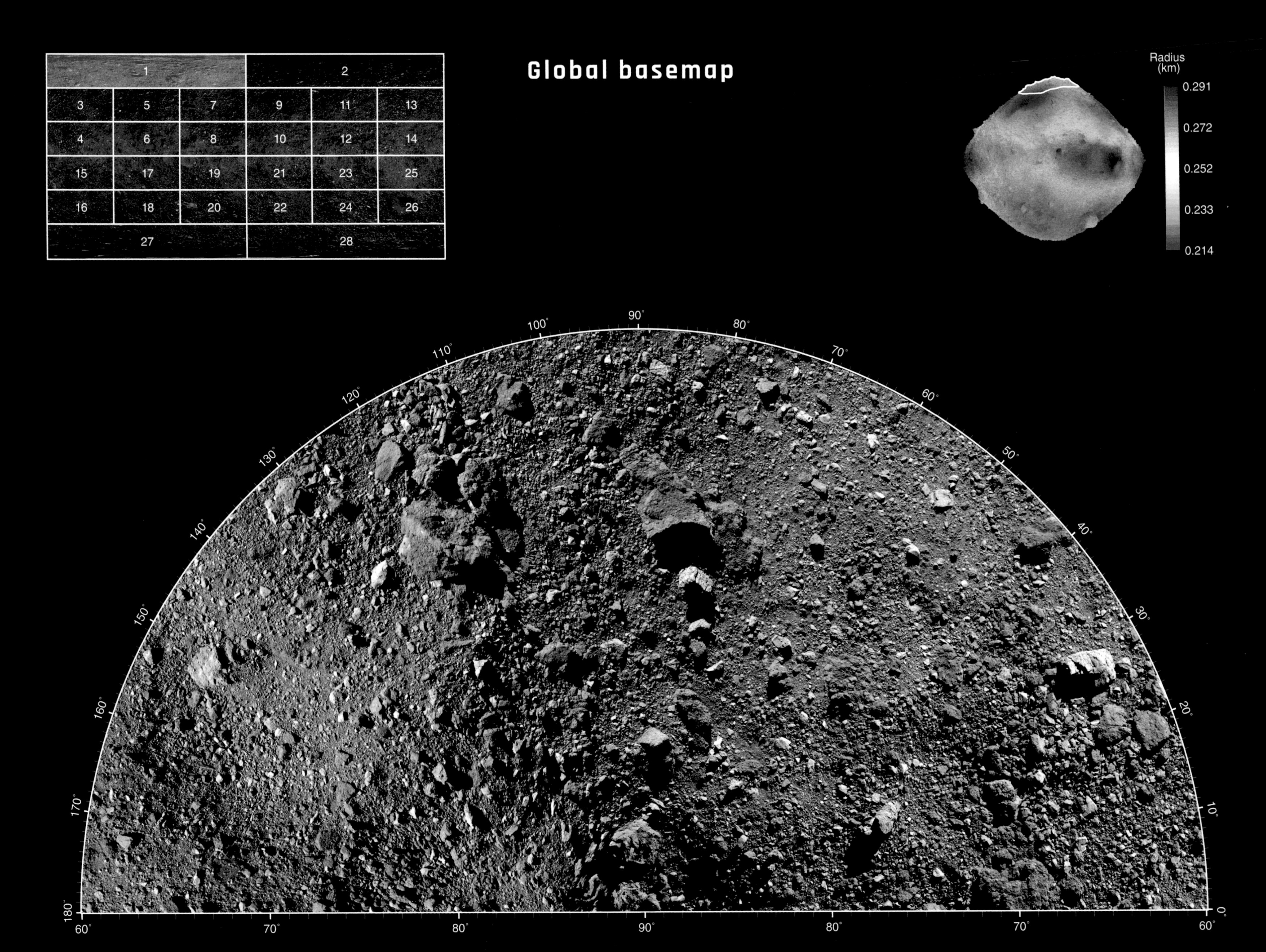
Global basemap
1
2
3
5
7
9
11
13
4
6
8
10
12
14
15
17
19
21
23
25
16
18
20
22
24
26
27
28
Radius
(km)
0.291
0.272
0.252
0.233
0.214
180°
170°
160°
150°
140°
130°
120°
110°
100°
90°
80°
70°
60°
50°
40°
30°
20°
10°
0°
60°
70°
80°
90°
80°
70°
60°

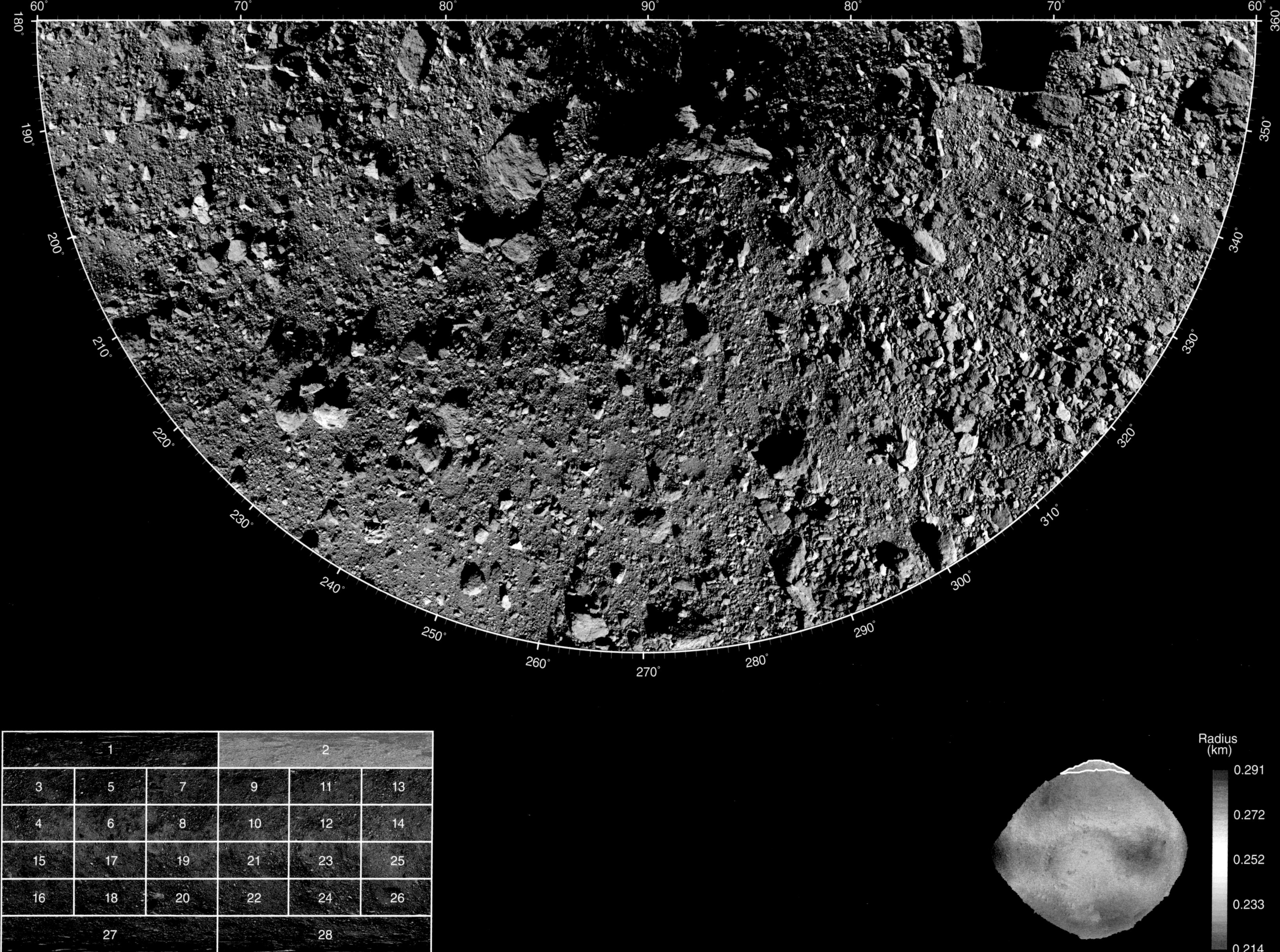

60°
70°
80°
90°
80°
70°
60°
180°
190°
200°
210°
220°
230°
240°
250°
260°
270°
280°
290°
300°
310°
320°
330°
340°
350°
360°
1
2
3
5
7
9
11
13
4
6
8
10
12
14
15
17
19
21
23
25
16
18
20
22
24
26
27
28
Radius
(km)
0.291
0.272
0.252
0.233
0.214

Global basemap
1
2
3
5
7
9
11
13
4
6
8
10
12
14
15
17
19
21
23
25
16
18
20
22
24
26
27
28
Radius
(km)
0.291
0.272
0.252
0.233
0.214
0°
10°
20°
30°
40°
50°
60°
60°
50°
40°
30°

30°
20°
10°
0°
0°
10°
20°
30°
40°
50°
1
2
3
5
7
9
11
13
4
6
8
10
12
14
15
17
19
21
23
25
16
18
20
22
24
26
27
28
Radius (km)
0.29
0.27
0.25
0.23
0.21

Global basemap
1
2
3
5
7
9
11
13
4
6
8
10
12
14
15
17
19
21
23
25
16
18
20
22
24
26
27
28
Radius (km)
0.29
0.27
0.25
0.23
0.21
60°
70°
80°
90°
100°
110°
60°
50°
40°
30°

20°
10°
0°
60°
70°
80°
90°
100°
110°
1
2
3
5
7
9
11
13
4
6
8
10
12
14
15
17
19
21
23
25
16
18
20
22
24
26
27
28
Radius
(km)
0.291
0.272
0.252
0.233

Global basemap
1
2
5
7
9
11
13
6
8
10
12
14
17
19
21
23
25
18
20
22
24
26
27
28
Radius (km)
0.29
0.27
0.25
0.23
0.21
130°
140°
150°
160°
170°

30°
20°
10°
0°
120°
130°
140°
150°
160°
170°
180°
1
2
3
5
7
9
11
13
4
6
8
10
12
14
15
17
19
21
23
25
16
18
20
22
24
26
27
28
Radius (km)
0.291
0.272
0.252
0.233
0.214

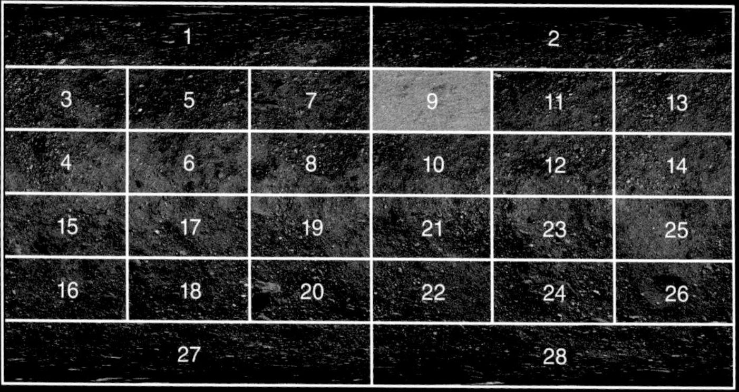

1 2
3 5 7 9 11 13
4 6 8 10 12 14
15 17 19 21 23 25
16 18 20 22 24 26
27 28

Global basemap

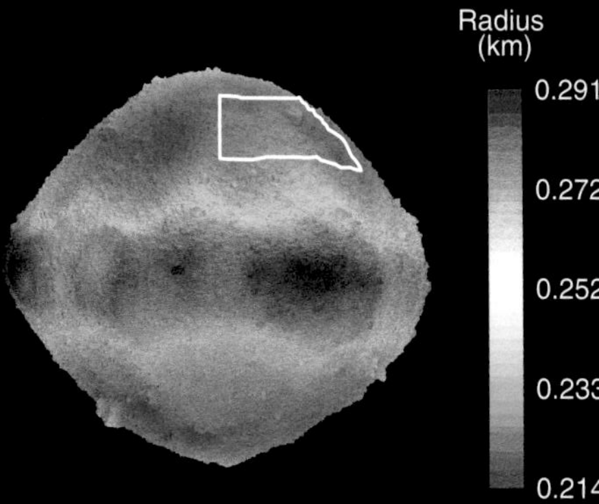

Radius (km)
0.291
0.272
0.252
0.233
0.214

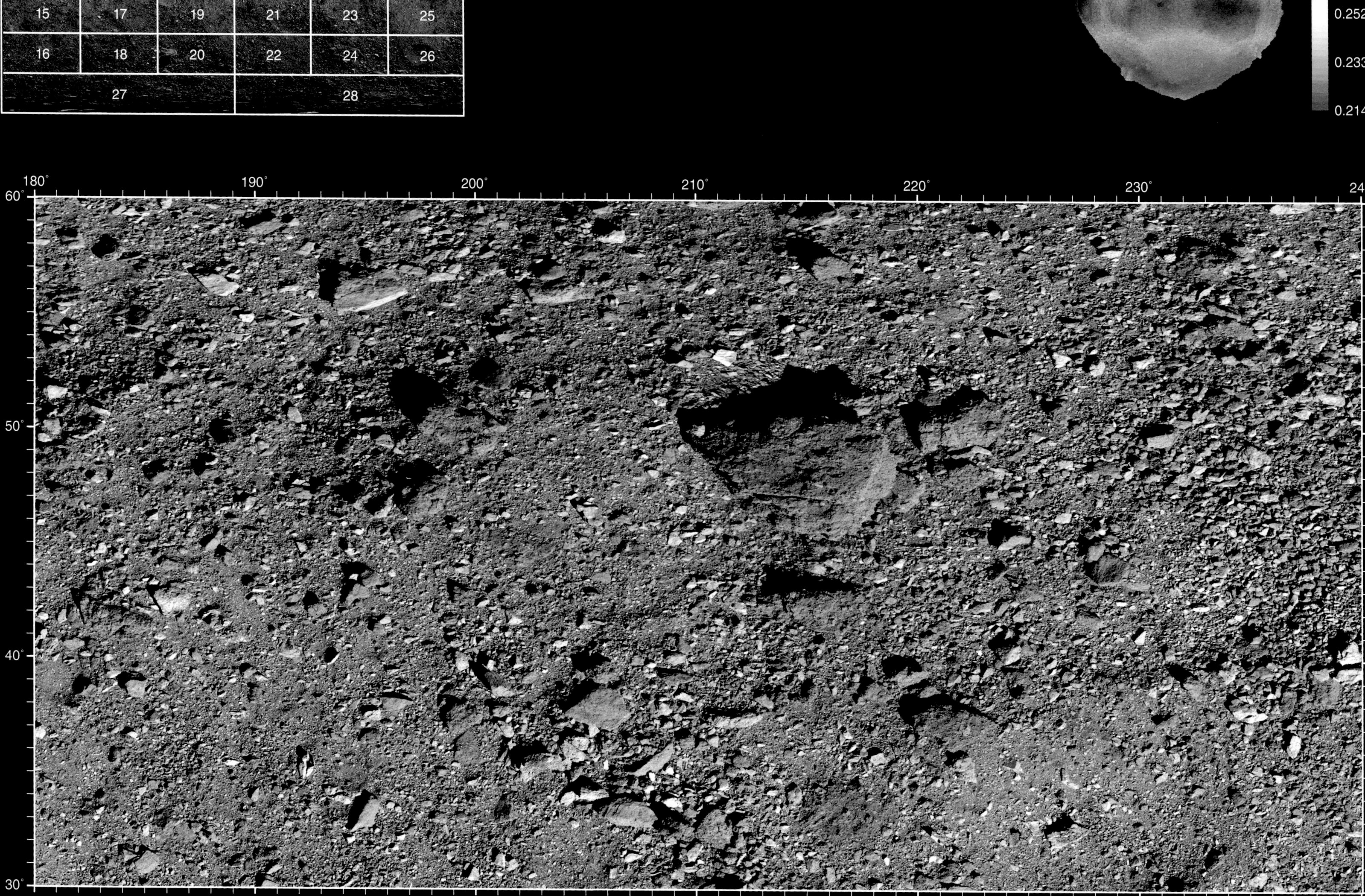

30°
20°
10°
0°
180°
190°
200°
210°
220°
230°
1
2
3
5
7
9
11
13
4
6
8
10
12
14
15
17
19
21
23
25
16
18
20
22
24
26
Radius (km)
0.29
0.27
0.25
0.23

3
5
7
9
11
13
4
6
8
10
12
14
5
17
19
21
23
25
6
18
20
22
24
26
27
28
0.27
0.25
0.23
0.2
0°
250°
260°
270°
280°
290°

30°
20°
10°
0°
240°
250°
260°
270°
280°
290°
300°
1
2
3
5
7
9
11
13
4
6
8
10
12
14
15
17
19
21
23
25
16
18
20
22
24
26
27
28
Radius (km)
0.291
0.272
0.252
0.233
0.214

Global basemap
1
2
3
5
7
9
11
13
4
6
8
10
12
14
15
17
19
21
23
25
16
18
20
22
24
26
27
28
Radius (km)
0.291
0.272
0.252
0.233
0.214
300°
310°
320°
330°
340°
350°
360
60°
50°
40°
30°

30°
20°
10°
0°
300°
310°
320°
330°
340°
350°
36
1
2
3
5
7
9
11
13
4
6
8
10
12
14
15
17
19
21
23
25
16
18
20
22
24
26
27
28
Radius
(km)
0.29
0.272
0.252
0.233
0.214

Global basemap

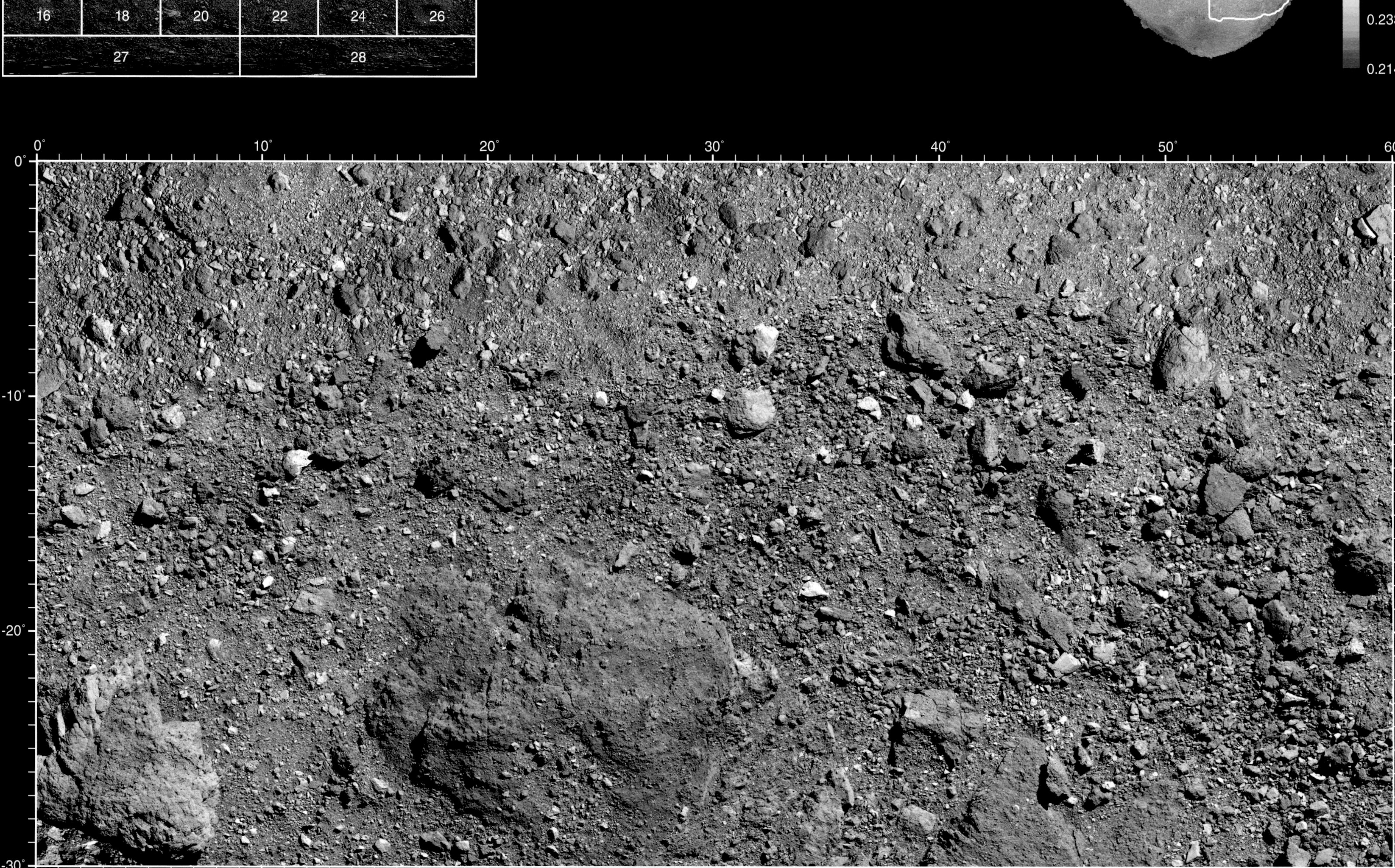

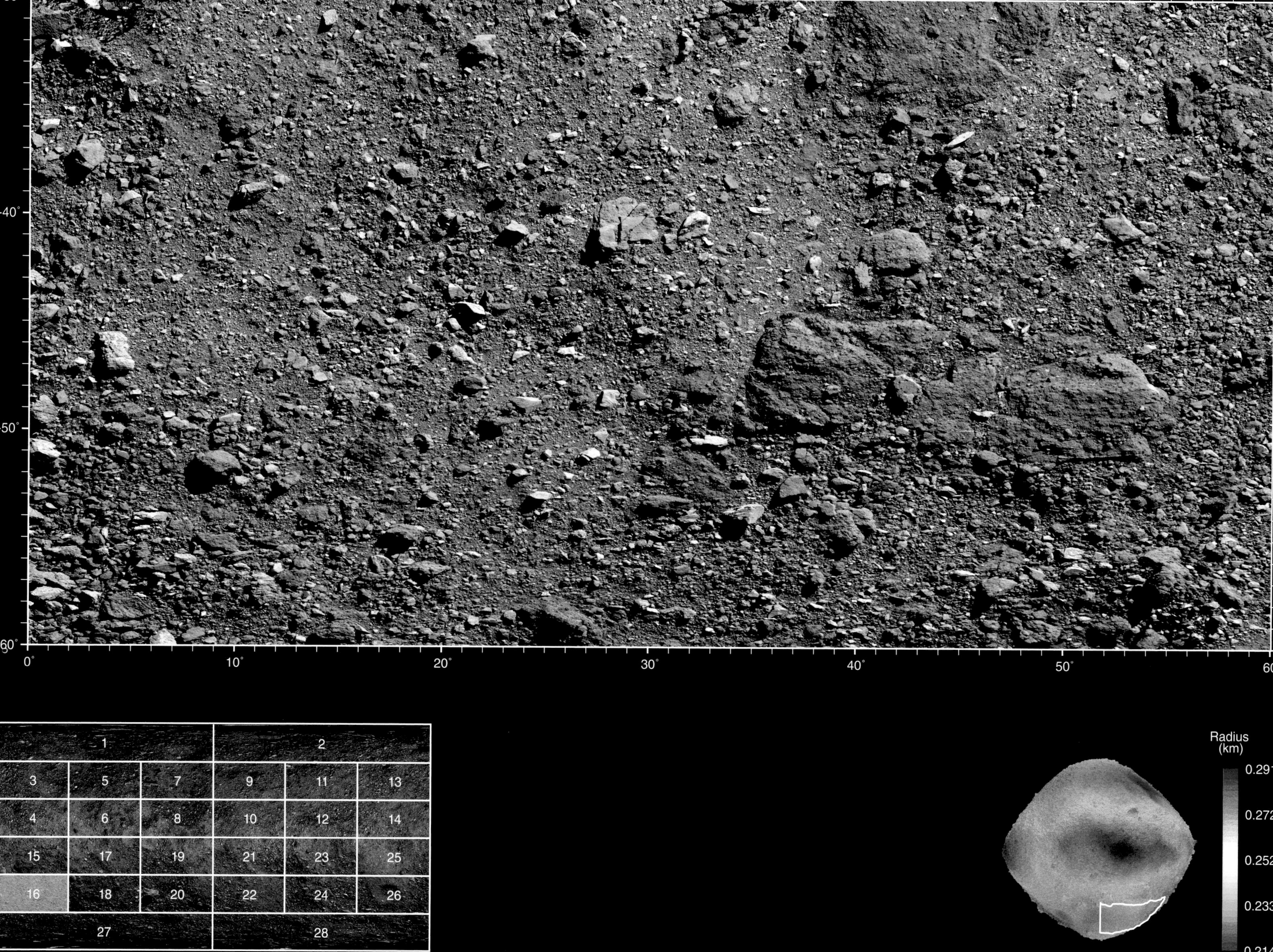
-40°
-50°
0°
10°
20°
30°
40°
50°
1
2
3
5
7
9
11
13
4
6
8
10
12
14
15
17
19
21
23
25
16
18
20
22
24
26
27
28
Radius (km)

Global basemap

1 2
3 5 7 9 11 13
4 6 8 10 12 14
15 17 19 21 23 25
16 18 20 22 24 26
27 28

Radius (km)

70° 80° 90° 100° 110°

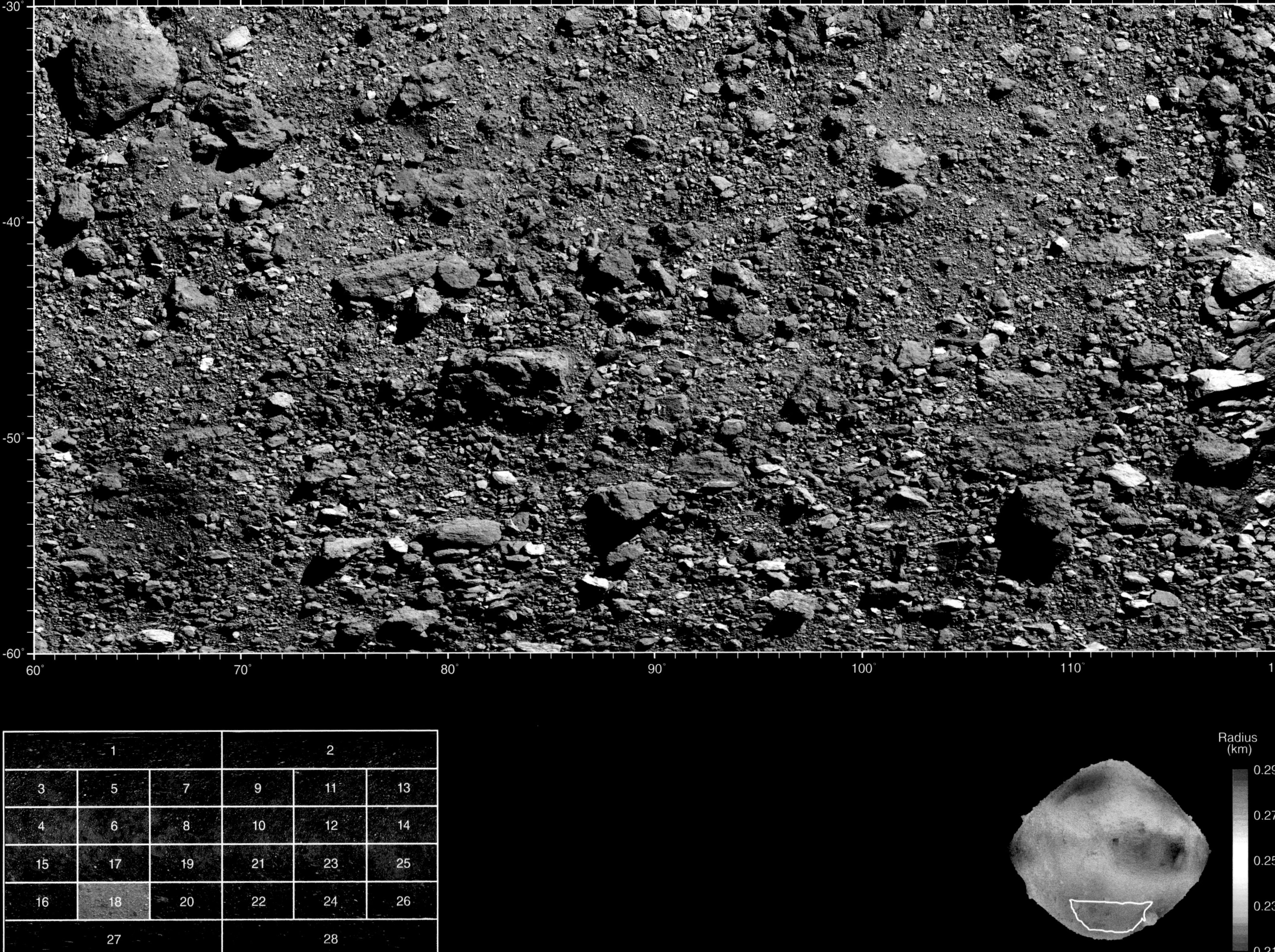
-30°
-40°
-50°
-60°
60°
70°
80°
90°
100°
110°
1
2
3
5
7
9
11
13
4
6
8
10
12
14
15
17
19
21
23
25
16
18
20
22
24
26
27
28
Radius (km)

Global basemap

1			2		
3	5	7	9	11	13
4	6	8	10	12	14
15	17	19	21	23	25
16	18	20	22	24	26
27			28		

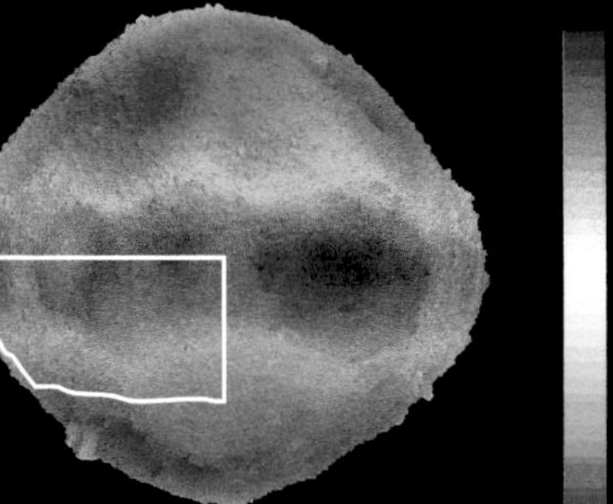

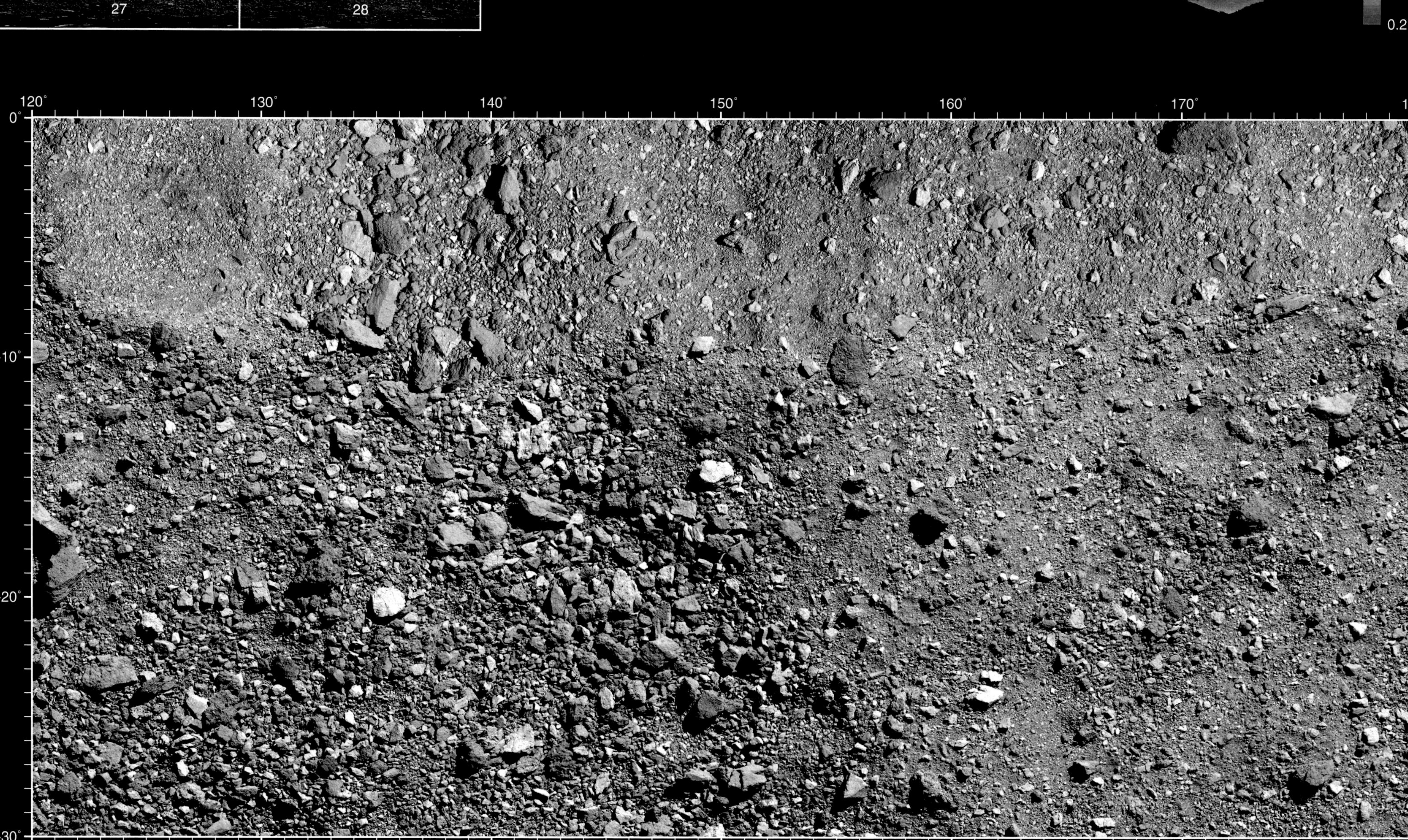

-30°
-40°
-50°
-60°
120°
130°
140°
150°
160°
170°
1
2
3
5
7
9
11
13
4
6
8
10
12
14
15
17
19
21
23
25
16
18
20
22
24
26
27
28
Radius
(km)
0.29
0.27
0.25
0.23
0.21

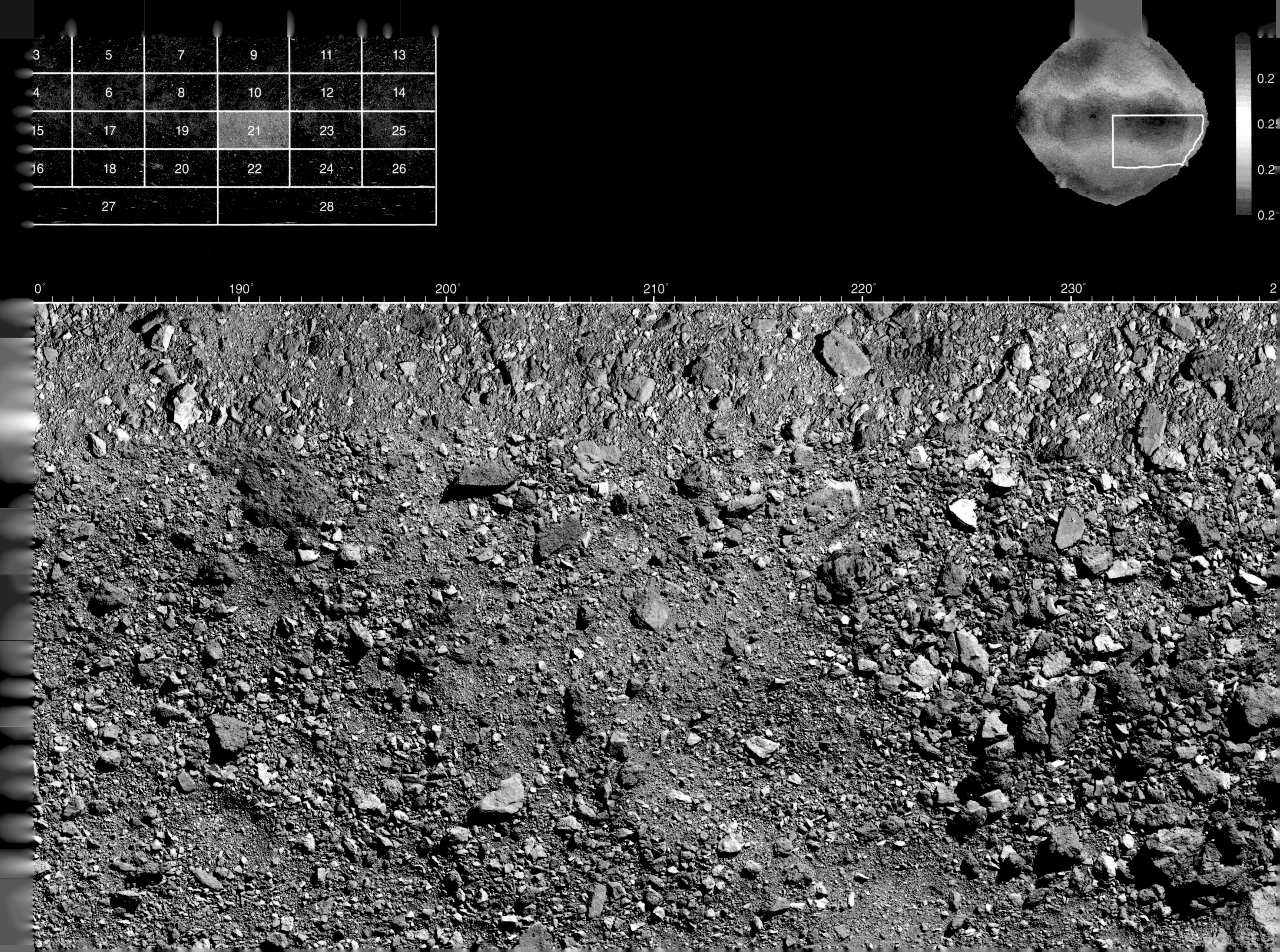
3
5
7
9
11
13
4
6
8
10
12
14
15
17
19
21
23
25
16
18
20
22
24
26
27
28
190°
200°
210°
220°
230°

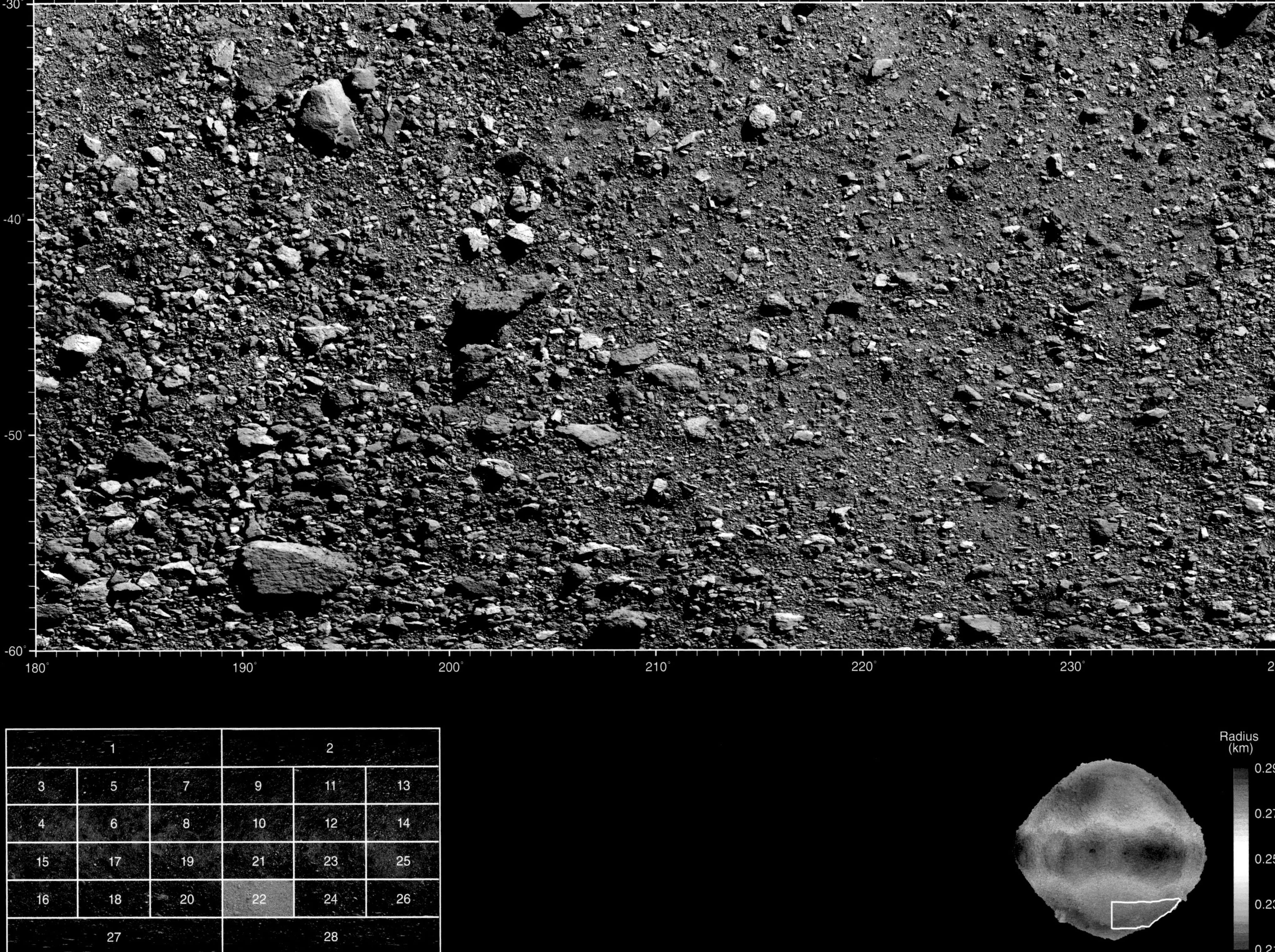

-30°
-40°
-50°
-60°
180°
190°
200°
210°
220°
230°
1
2
3
5
7
9
11
13
4
6
8
10
12
14
15
17
19
21
23
25
16
18
20
22
24
26
27
28
Radius
(km)

Global basemap
1
2
3
5
7
9
11
13
4
6
8
10
12
14
15
17
19
21
23
25
16
18
20
22
24
26
27
28
Radius
(km)
0.29
0.27
0.25
0.23
0.21
240°
250°
260°
270°
280°
290°
0°
-10°
-20°
-30°

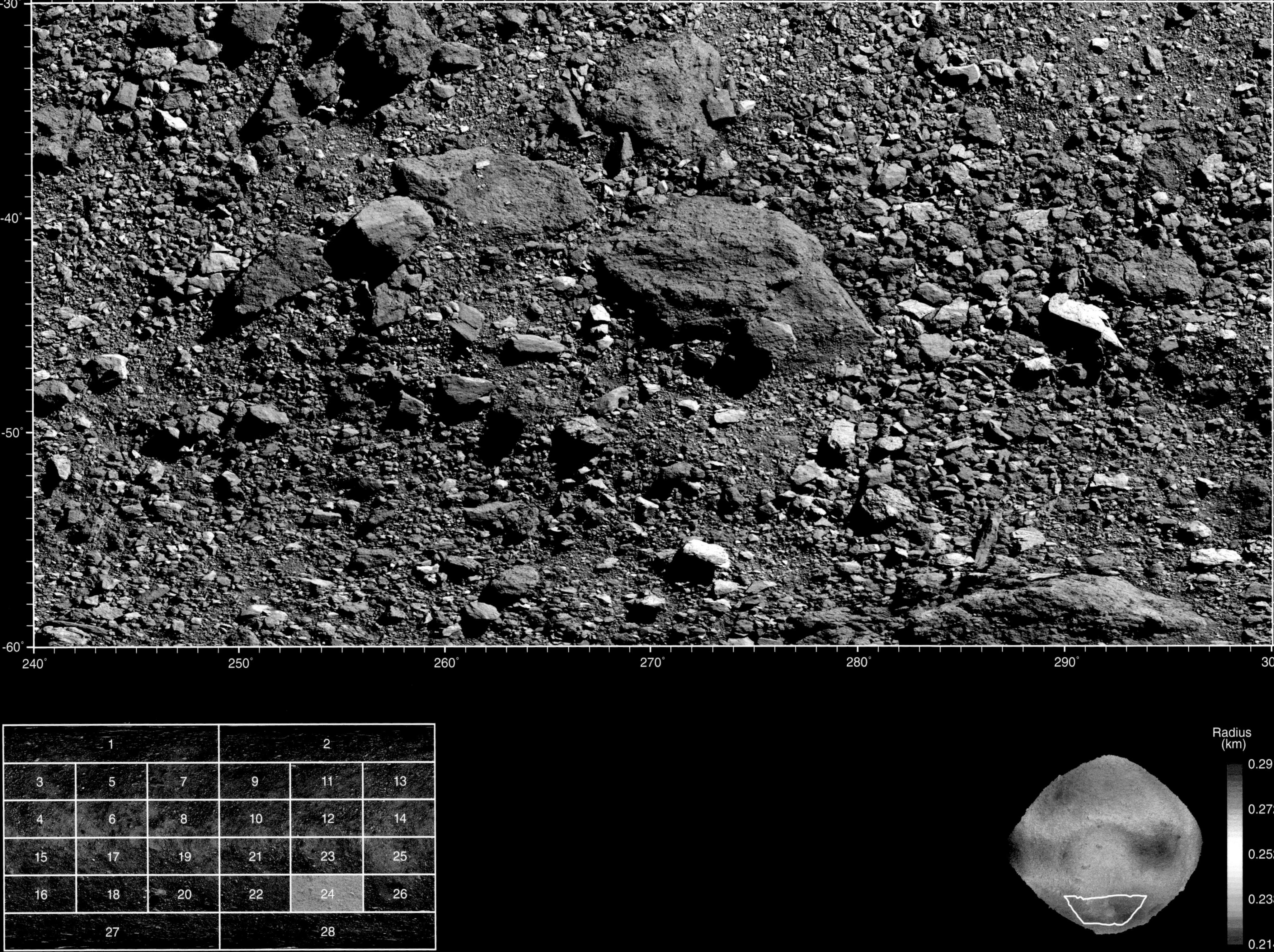

-30°
-40°
-50°
-60°
240°
250°
260°
270°
280°
290°
1
2
3
5
7
9
11
13
4
6
8
10
12
14
15
17
19
21
23
25
16
18
20
22
24
26
27
28
Radius (km)
0.29
0.27
0.25
0.23
0.21

Global basemap

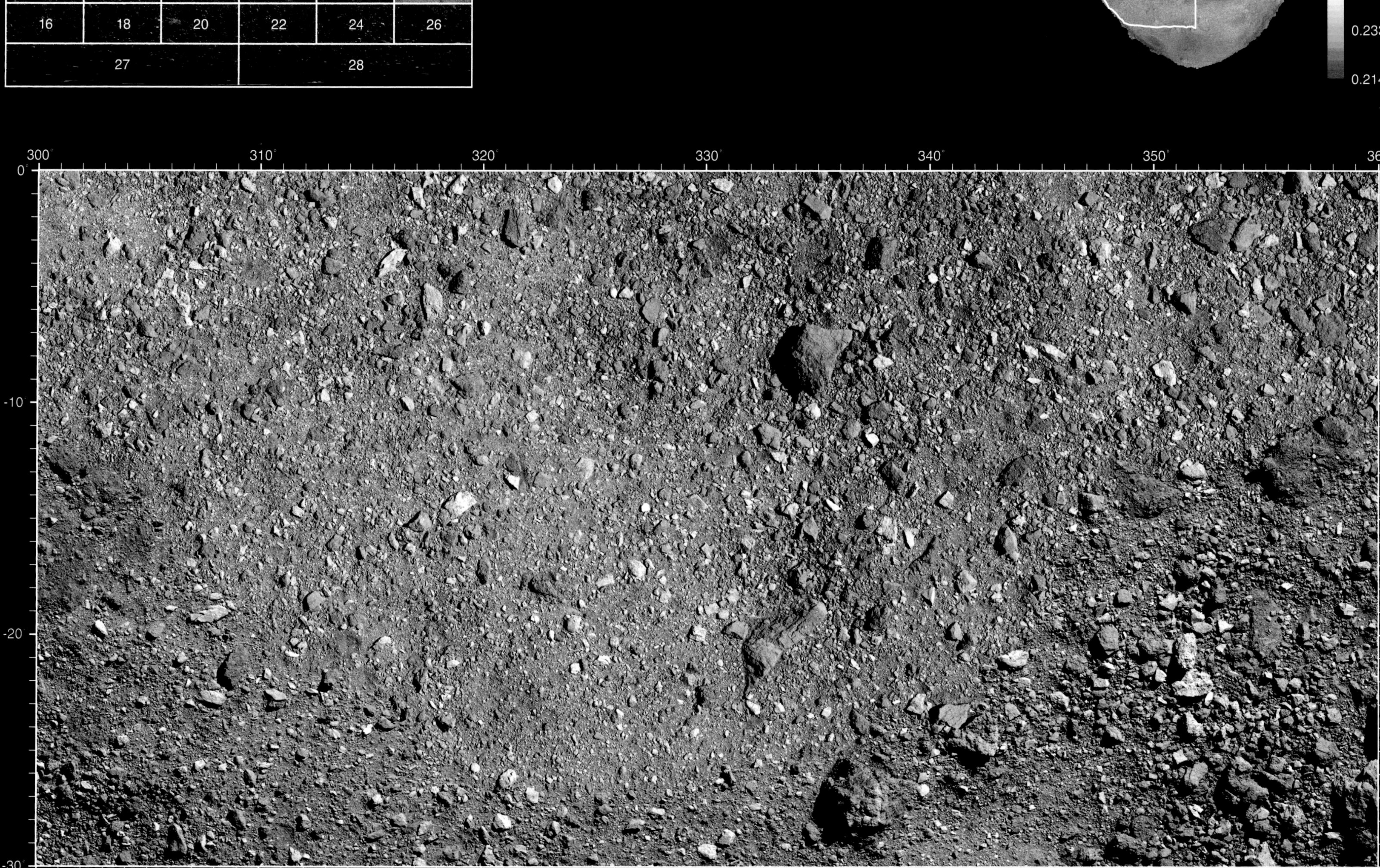

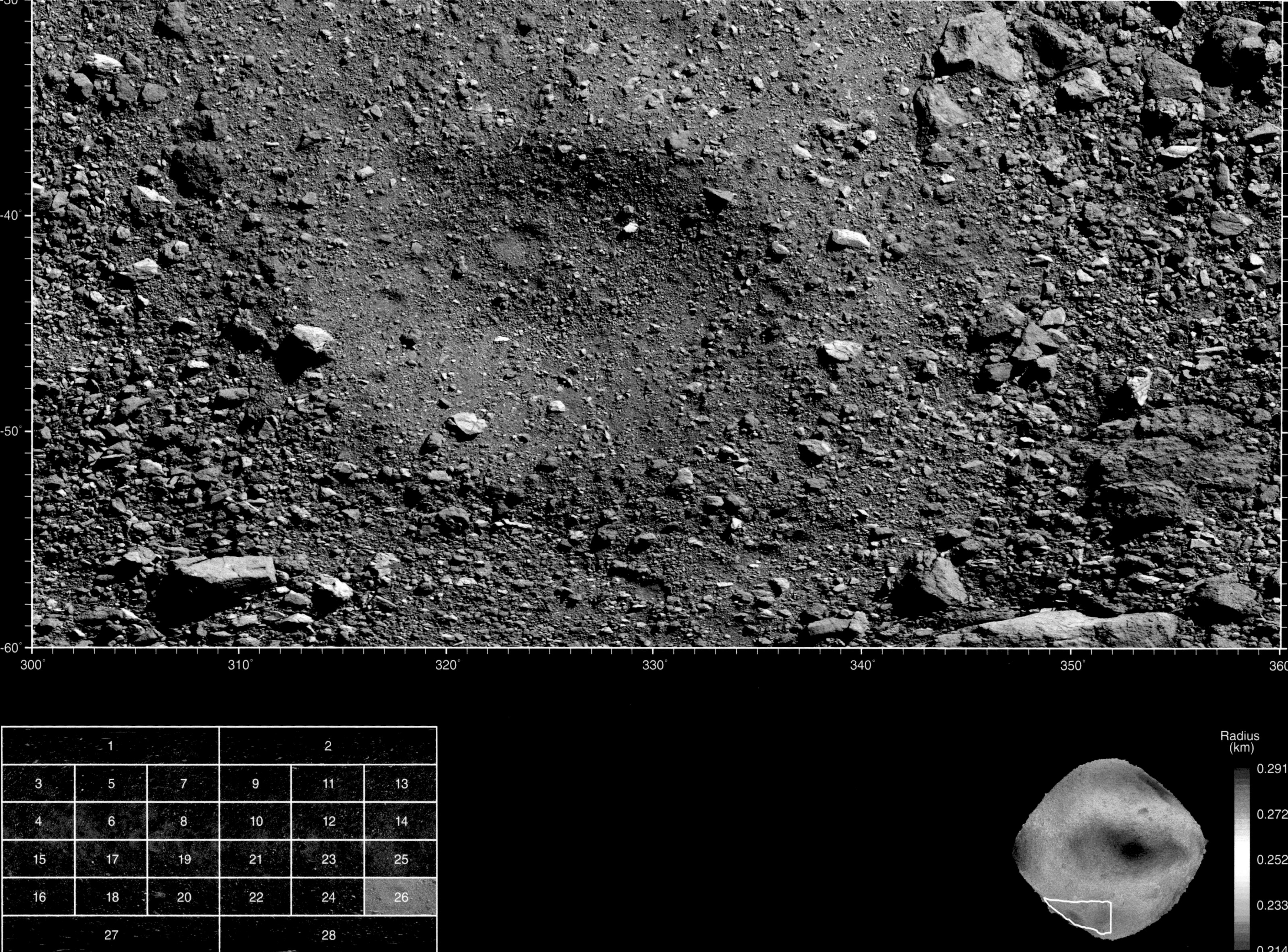

-40°
-50°
-60°
300°
310°
320°
330°
340°
350°
360°
1
2
3
5
7
9
11
13
4
6
8
10
12
14
15
17
19
21
23
25
16
18
20
22
24
26
27
28
Radius (km)
0.291
0.272
0.252
0.233

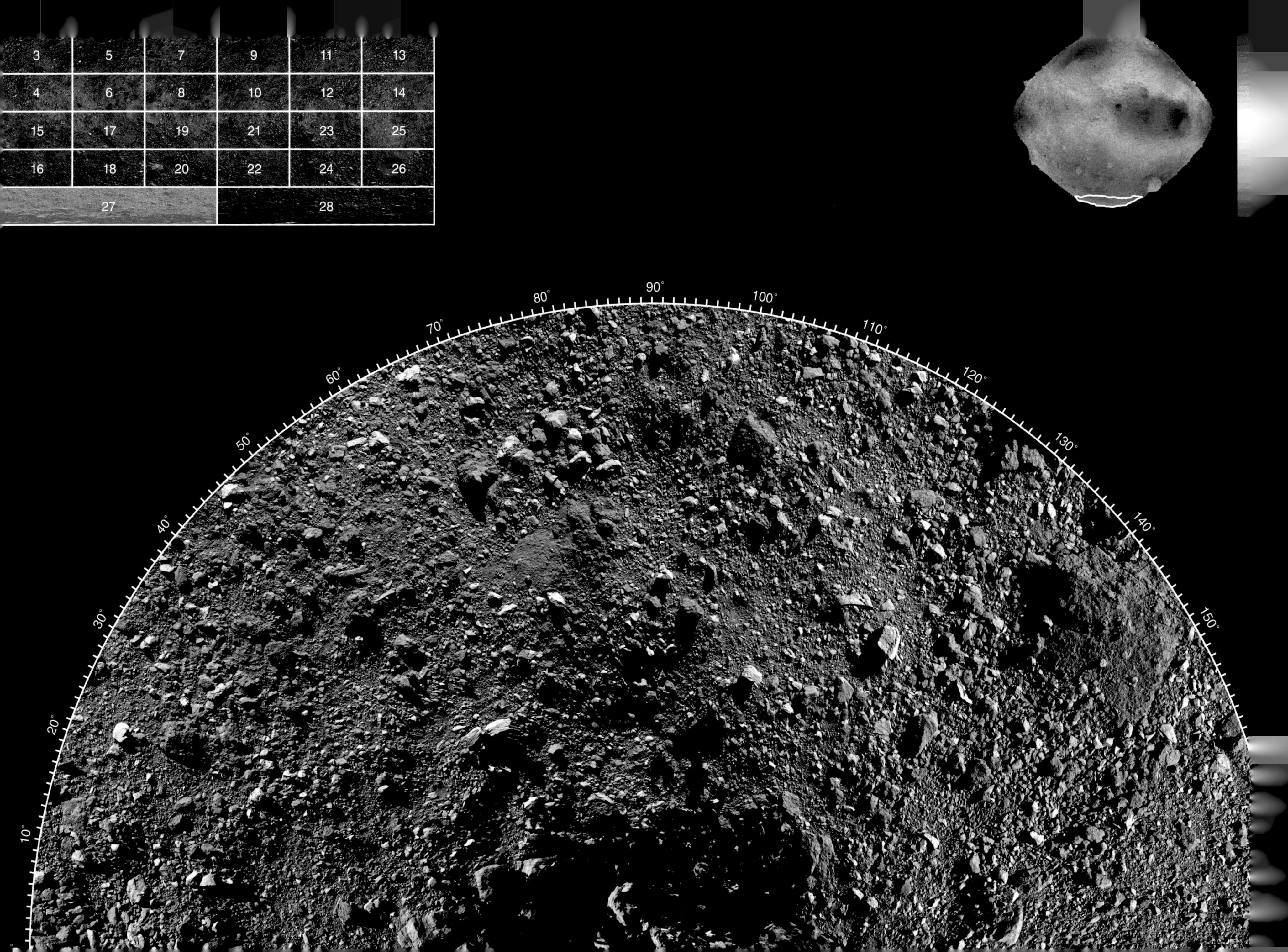

3
5
7
9
11
13
4
6
8
10
12
14
15
17
19
21
23
25
16
18
20
22
24
26
27
28
10°
20°
30°
40°
50°
60°
70°
80°
90°
100°
110°
120°
130°
140°
150°

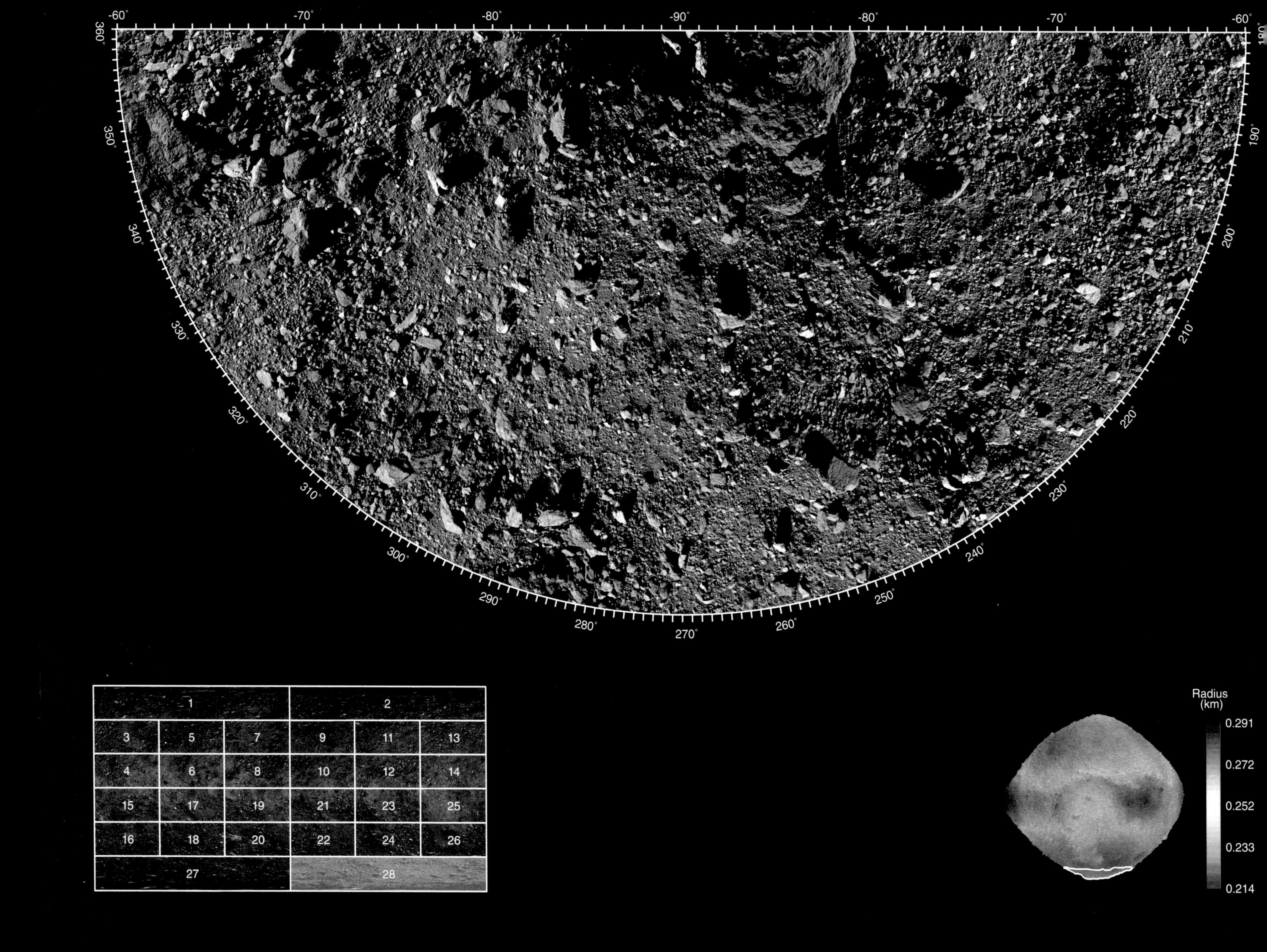
-60°
-70°
-80°
-90°
-80°
-70°
-60°
360°
350°
340°
330°
320°
310°
300°
290°
280°
270°
260°
250°
240°
230°
220°
210°
200°
190°
180°
1
2
3
5
7
9
11
13
4
6
8
10
12
14
15
17
19
21
23
25
16
18
20
22
24
26
27
28
Radius (km)
0.291
0.272
0.252
0.233
0.214

Features index

Landscape features on Bennu with official IAU names (as of printing) are tabulated alphabetically below, accompanied by their center coordinates, size, and thumbnail image. In the case of boulders, "diameter" refers to the longest visible axis. We report values measured by the OSIRIS-REx science team, which differ modestly in some instances from those defined by the IAU in the Gazetteer of Planetary Nomenclature. Thumbnail images are from the global basemap in Chapter 7.

Feature name	Center latitude (degrees)	Center longitude (degrees)	Diameter (meters)	Thumbnail
Aellopus Saxum	25.44	335.67	5.7	
Aetos Saxum	3.46	150.36	9.1	
Alicanto Crater	−0.14	111.23	76	
Amihan Saxum	−17.96	256.51	33	
Benben Saxum	−45.86	127.59	52	

Boobrie Saxum	48.08	214.28	30	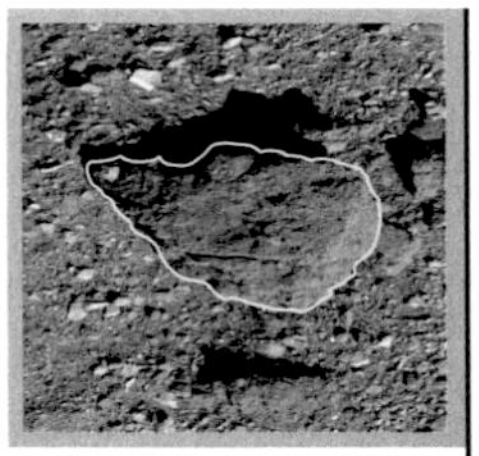
Bralgah Crater	−43.28	325.20	63	
Camulatz Saxum	−10.26	259.65	22	
Celaeno Saxum	18.42	335.23	7.7	
Ciinkwia Saxum	−4.97	249.47	21	
Dinewan Crater	−41.51	323.41	8.4	
Dodo Saxum	−32.68	64.42	27	
Gamayun Saxum	9.86	105.45	19	

Gargoyle Saxum	4.59	92.48	18	
Gullinkambi Saxum	18.53	17.96	31	
Hokioi Crater	54.90	41.70	24	
Huginn Saxum	−29.77	43.25	46	
Huhuk Crater	−1.24	152.61	83	
Kongamato Saxum	5.03	66.31	19	
Lilitu Crater	−27.04	114.80	28	
Minokawa Crater	−7.73	269.20	187	

Muninn Saxum	−29.34	48.68	28	
Ocypete Saxum	25.09	328.25	5.6	
Odette Saxum	−44.86	291.08	19	
Odile Saxum	−42.74	294.08	25	
Ohnivak Crater	−3.74	126.81	56	
Pegasus Crater	23.55	329.88	12	
Pouakai Saxum	−40.45	166.75	10.6	
Roc Saxum	−23.46	25.36	96	

Sampati Crater	−19.02	97.32	11	
Simurgh Saxum	−25.32	4.05	39	
Strix Saxum	13.40	88.26	5.3	
Thorondor Saxum	−47.94	45.10	61	
Tlanuwa Regio	−37.86	261.70	301	
Wuchowsen Crater	11.77	87.99	21	
Wututu Crater	−16.93	84.28	14	

Bennu by numbers

Tabulated here are the best measurements and derivations of Bennu's dynamical, physical, optical, and spectroscopic properties that were available ahead of sample return. The reported uncertainties are ±1 standard deviation. (Uncertainty is a measure of how well a value is known; smaller is better.)

Orbit	
Epoch	2455562.5 (2011-Jan-01.0) Barycentric Dynamical Time Reference: JPL 118 (heliocentric IAU76/J2000 ecliptic)
Orbital inclination (degrees)	6.03494377025 ± 0.00000000686
Orbital eccentricity	0.2037450762416 ± 0.0000000000697
Obliquity (degrees)	177.54 ± 0.01
Aphelion (astronomical units)	1.35588765134 ± 0.00000000024
Perihelion (astronomical units)	0.896894400446 ± 0.00000000023
Time of perihelion passage (Barycentric Dynamical Time)	2010-08-30.6419408727 ± 0.0000000225
Longitude of ascending node (degrees)	2.0608661957 ± 0.0000000556
Argument of perihelion (degrees)	66.2230608408 ± 0.0000000644
Mean anomaly (degrees)	101.703952002457 ± 0.00000004275
Orbital period (Earth days)	436.64872811 ± 0.00000011
Change in semi-major axis (meters per year)	-284.6 ± 0.2

Rotation	
Sidereal rotation period (epoch of February 1, 2019) (hours)	4.296005229 ± 0.000000019
Rotational acceleration (seconds per century)	-1.02 ± 0.15
Pole position (right ascension, declination) (J2000; degrees)	(85.45218 ± 0.00034), (-60.36780 ± 0.00010)
Center-of-mass/center-of-figure offset (x, y, z) (meters)	(1.31 ± 0.03, -0.46 ± 0.04, 0.22 ± 0.01)
Products of inertia (assuming constant density) (square meters)	I_{zx} = -46.70 ± 0.05 I_{zy} = 11.39 ± 0.01

Size, shape, & density	
Average radius (meters)	242.22 ± 0.15
Best-fit ellipsoid (meters)	(252.37 ± 0.09) x (245.91 ± 0.09) x (228.37 ± 0.09)
Surface area (square kilometers)	0.78740 ± 0.0004
Volume (cubic kilometers)	0.061354 ± 0.00006
Mass (1010 kilograms)	7.327 ± 0.003
Bulk density (kilograms per cubic meter)	1194 ± 3

Terrain	
Average surface roughness (degrees)	40 ± 3
Average surface slope, not including the contribution of boulders (degrees)	22
Average surface slope, including the contribution of boulders (degrees)	33
Range of particle sizes	sub-micron (dust) to ~100 meters
Power-law index of particle size frequency distribution	2.9 ± 0.3
Gravitational & thermal properties	
GM (cubic meters per square second)	4.890450 ± 0.0009
Average gravitational acceleration (meters per square second)	0.0000542 ± 0.0000001
Hill sphere radius (kilometers)	31.05 ± 0.01
Average thermal inertia (Joules meter^{-2} Kelvin^{-1} second$^{-½}$)	300 ± 30

Gravitational & thermal properties	
GM (cubic meters per square second)	4.890450 ± 0.0009
Average gravitational acceleration (meters per square second)	0.0000542 ± 0.0000001
Hill sphere radius (kilometers)	31.05 ± 0.01
Average thermal inertia (Joules meter^{-2} Kelvin^{-1} second$^{-1/2}$)	300 ± 30

Optical & spectroscopic properties	
Peak brightnesses during close approaches to Earth (magnitude)	14.4 (1999–2000), 16.1 (2005–2006), 19.9 (2011–2012)
Global normal albedo (radiance factor, I/F)	0.046 ± 0.002
Normal albedo range of darkest boulders (I/F)	0.033–0.037
Normal albedo range of brightest boulders (I/F)	0.1–0.3
Average spectral slope from wavelengths 0.44 to 0.89 μm (per micron)	−0.1701
Spectral type	B
Meteorite spectral analogs	CI, CM, CR

Mission timeline

Tabulated here are the dates of mission phases (uppercase) and key spacecraft activities (lowercase) during proximity operations at Bennu. The Recon and Rehearsal phases partially overlapped one another.

Phase / activity	Start date	End date
APPROACH	17 August 2018	3 December 2018
Arrival & PRELIMINARY SURVEY	3 December 2018	31 December 2018
ORBITAL A	31 December 2018	22 February 2019
DETAILED SURVEY	22 February 2019	12 June 2019
Flyby operations	7 March 2019	19 April 2019
Refly of missed Flyby 2	26 September 2019	
Equatorial Stations observations	25 April 2019	6 June 2019
ORBITAL B	12 June 2019	6 August 2019
ORBITAL C	6 August 2019	9 September 2019
RECONNAISSANCE (RECON)	9 September 2019	1 June 2020
Recon A flybys	5 October 2019	26 October 2019
Orbital R (safe home orbit during Recon)	31 October 2019	6 January 2020
Recon B sorties	21 January 2020	11 February 2020
Recon C sorties	3 March 2020	26 May 2020
REHEARSALS & SAMPLE COLLECTION	30 March 2020	28 October 2020
Checkpoint Rehearsal	14 April 2020	
Matchpoint Rehearsal	11 August 2020	
Sample Collection (TAG)	20 October 2020	
Sample Stow	28 October 2020	
Final Flyby	7 April 2021	
Departure	10 May 2021	

Image credits

All photographs taken by the OSIRIS-REx Camera Suite (PolyCam, MapCam, and SamCam) – NASA/Goddard/University of Arizona

All photographs taken by the OSIRIS-REx Touch-and-Go Camera System (NavCams and StowCam) – NASA/Goddard/University of Arizona/Lockheed Martin

All stereo images created by Claudia Manzoni and Brian May.

All computer-generated visualizations of the spacecraft – NASA/GSFC/SVS

Page 6, Professor Lauretta with the OSIRIS-REx spacecraft – Lockheed Martin.

Chapter 1

Image of Allan Hills meteorite thin section – D. Lauretta, from Lauretta & Kilgore (2005).
Images from the Dawn mission – NASA/JPL/Caltech/UCLA/MPS/DLR/IDA
Images from the NEAR Shoemaker mission – NASA/JPL/JHUAPL
Images from the Hayabusa mission – JAXA/ISAS
Images from the VLT – ESO

Photos of Murchison and Allende meteorites – University of Arizona/D. Hill
Photos of ANSMET in Antarctica – D. Lauretta (far left, middle left, middle right) & D. Glavin (far right)
Photos of Lonewolf Nunataks meteorite – NASA

Credits for timeline images:
If not detailed below, other images NASA/Goddard/UA or James Symonds.
1492 Ensisheim Meteor Storm – Wikimedia Creative Commons/H.Raab
1794 Ernst Chladni – Public Domain
1801 Gieuseppe Piazzi – INAF – Osservatorio di Palermo/Gieuseppe S. Vaiana
1802 Edward Howard – public domain
1803 L'Aigle Meteorite – public domain
1945 Meteor – public domain
1968 Icarus – Goldstone Observatory NASA/JPL-Caltech/GSSR
1995 Spacewatch telescope, Kitt's Peak – NOIRLab/AURA/NSF
2004 Stewart Observatory – Catalina Sky Survey
2008/9 Rosetta spacecraft – ESA
2010 Pan-STARRS Observatory, Hawaii – R. Ratkowski
2010/15 Dawn spacecraft – NASA/JPL
2019/22 Hayabusa2 – JAXA

Chapter 2

Photos of OSIRIS-REx launch and detach – United Launch Alliance
Photos of the spacecraft and TAGSAM during testing and preparation – Lockheed Martin
Stills of the OCAMS imagers in the laboratory – University of Arizona/S. Platts
Bennu target selection pyramid – adapted from Lauretta et al. (2017)
Mission phase diagrams – adapted from Lauretta et al. (2021)

Chapter 3

Discovery images of Bennu – Klet Observatory/J. Tichá & M. Tichý
Radar images of Bennu – JPL/Caltech
Radar shape model of Bennu – adapted from Nolan et al. (2013)
Bennu orbital diagram – adapted from Lauretta et al. (2017)
Ejected particle trajectories – adapted from M. Brozović/JPL/Caltech/NASA, jpl.nasa.gov/images/pia24101-asteroid-bennus-particle-ejection-events-animation; data from Chesley et al. (2020)

Chapter 6

Photo of E. Church and team at mission control – NASA/GSFC
Spectral map of Nightingale before and after sampling – adapted from Lauretta et al. (2022)
Illustration of OSIRIS-REx releasing the return capsule over Earth – NASA/GSFC

Stereoscopic picture credits

NASA/Goddard/University of Arizona/CM-BM with the exception of:

p.11: The Brian May Archive of Stereoscopy
p.19: NASA/JPL-CalTech/UCLA/MPS/DLR/IDA/CM-BM (top & middle), ESO/M. Marsset et al./MISTRAL algorithm (ONERA/CNRS)/CM-BM (bottom)
p.26: NASA/JPL/JHUAPL/BM (top), ISAS/JAXA/CM-BM (middle), JAXA/University of Tokyo/ Kochi University/Rikkyo University/Nagoya University/Chiba Institute of Technology/Meiji University/University of Aizu/AIST/CM-BM (bottom)
p.31: NASA/These 3D reconstructed image data were produced at the Antarctic Meteorite Sample Laboratory Facility for Astromaterials 3D in NASA's Acquisition & Curation Office and were funded by NASA Planetary Data Archiving, Restoration, and Tools Program, Proposal No.: 15-PDART15_2-0041/CM-BM
p.34: NASA/Sandy Joseph and Tim Terry/CM-BM
p.35: JAXA/University of Tokyo/ Kochi University/Rikkyo University/Nagoya University/Chiba Institute of Technology/Meiji University/University of Aizu/AIST/CM-BM (Ryugu, bottom)
p.41: NASA's Goddard Space Flight Center/CM-BM
p.47: NASA's Goddard Space Flight Center/Scientific Visualization Studio/BM
p.53: Klet Observatory/J. Tichá & M. Tichý/BM
p.54: NASA/Goddard/University of Arizona/DL-BM
p.67: M. Brozovic/JPL/Caltech/NASA/University of Arizona/CM-BM
p.129: NASA's Goddard Space Flight Center/CM-BM
p.134 NASA/Goddard/University of Arizona/Lockheed Martin/CM-BM (top), NASA's Goddard Space Flight Center/CI Lab/SVS/CM-BM (bottom)
p.138: NASA's Goddard Space Flight Center/CI Lab/SVS/CM-BM
p.140: NASA's Goddard Space Flight Center/CM-BM
(abbreviations: BM = Brian May, CM = Claudia Manzoni, DL = Dante Lauretta)

References

Adam, C.D. et al. (2022). Concept of Operations for OSIRIS-REx Optical Navigation Image Planning. AIAA SciTech Forum, San Diego, CA, 3–7 January 2022, doi:10.2514/6.2022-1569.

Antoniadi, E.M. (1939). On Ancient Meteorites, and the Origin of the Crescent and Star Emblem (with Plates VII, VIII). *J. R. Astron. Soc. Canada* 33, 177.

Antreasian, P. et al. (2022). OSIRIS-REx Proximity Operations and Navigation Performance at (101955) Bennu. AIAA SciTech Forum, San Diego, CA, 3–7 January 2022, doi:10.2514/6.2022-2470.

Archinal, B.A. et al. (2018). Report of the IAU Working Group on Cartographic Coordinates and Rotational Elements: 2015. *Celest. Mech. Dyn. Astr.* 130, 22.

Ballouz, R.-L. et al. (2020). Bennu's near-Earth lifetime of 1.75 million years inferred from craters on its boulders. *Nature* 587, 205–209.

Barnouin, O.S. et al. (2019). Shape of (101955) Bennu indicative of a rubble pile with internal stiffness. *Nat. Geosci.* 12, 247–252.

Barnouin, O.S. et al. (2020). Digital terrain mapping by the OSIRIS-REx mission. *Planet. Space Sci.* 180, 104764.

Barnouin, O. & Nolan, M. (2021). Bennu Coordinate System Description, version 5.0. Origins, Spectral Interpretation, Resource Identification, Security, Regolith Explorer (OSIRIS-REx): Mission Bundle, Crombie, M.K. & Selznick, S., Eds., NASA Planetary Data System, urn:nasa:pds:orex.mission.

Barnouin, O.S. et al. (2022). Geologic Context of the OSIRIS-REx Sample Site from High-resolution Topography and Imaging. *Planet. Sci. J.* 3, 75.

Belton, M.J.S. et al. (1996). Galileo's encounter with 243 Ida: An overview of the imaging experiment. *Icarus* 120, 1–19.

Bennett, C.A. et al. (2021). A high-resolution global basemap of (101955) Bennu. *Icarus* 357, 113690.

Berry, K. et al. (2020). Revisiting OSIRIS-REx Touch-and-Go (TAG) Performance Given the Realities of Asteroid Bennu. *Proceedings of the 43rd Annual AAS GNC Conference*, AAS 20-088.

Bierhaus, E.B. et al. (2018). The OSIRIS-REx Spacecraft and the Touch-and-Go Sample Acquisition Mechanism (TAGSAM). *Space Sci. Rev.* 214, 107.

Bierhaus, E.B. et al. (2022). Crater population on asteroid (101955) Bennu indicates impact armouring and a young surface. *Nat. Geosci.* 15, 440–446.

Blau, P.J. et al. (1973). Investigation of the Canyon Diablo metallic spheroids and their relationship to the breakup of the Canyon Diablo meteorite. *J. Geophys. Res.* 78, 363–374.

Bos, B.J. et al. (2018). Touch And Go Camera System (TAGCAMS) for the OSIRIS-REx Asteroid Sample Return Mission. *Space Sci. Rev.* 214, 37.

Bos, B.J. et al. (2020). In-Flight Calibration and Performance of the OSIRIS-REx Touch And Go Camera System (TAGCAMS). *Space Sci. Rev.* 216, 71.

Bottke, W.F. et al. (2006). The Yarkovsky and YORP Effects: Implications for Asteroid Dynamics. *Annu. Rev. Earth Planet. Sci.* 34, 157–191.

Bottke, W.F. et al. (2015). In search of the source of asteroid (101955) Bennu: Applications of the stochastic YORP mode. *Icarus* 247, 191–217.

Bottke W.F. et al. (2020). Interpreting the Cratering Histories of Bennu, Ryugu, and Other Spacecraft-explored Asteroids. *Astron. J.* 160, 14.

Bottke, W.F. et al. (2020). Meteoroid Impacts as a Source of Bennu's Particle Ejection Events. *J. Geophys. Res. Planet.* 125, e2019JE006282.

Brearley, A.J. (2006). The action of water. In *Meteorites and the Early Solar System II*, Lauretta, D. S. et al., Eds. (Univ. Arizona Press), pp. 587–624.

Brewster, D. (1811). Supplementary Chapters. In *Ferguson's Astronomy* (Ballantyne and Co.), pp. 121–133.

Brouwer, D. (1963). The problem of the Kirkwood gaps in the asteroid belt. *Astron. J.* 68, 152.

Buchner, E. et al. (2012). Buddha from space – An ancient object of art made of a Chinga iron meteorite fragment. *Meteorit. Planet. Sci.* 47, 1491–1501.

Burke, J.G. (1991). *Cosmic Debris: Meteorites in History* (Univ. California Press).

Burke, K.N. et al. (2021). Particle Size-Frequency Distributions of the OSIRIS-REx Candidate Sample Sites on Asteroid (101955) Bennu. *Remote Sens.* 13, 1315.

Cambioni, S. et al. (2021). Fine-regolith production on asteroids controlled by rock porosity. *Nature* 598, 49–52.

Cameron, A.G.W. (1962). Formation of the solar nebula. *Icarus* 1, 339–342.

Campins, H. et al. (2010). The origin of asteroid 101955 (1999 RQ36). *Astrophs. J. Lett.* 721, L53.

Cassini, G.D. (1685). *Découverte de la Lumière Celeste Qui Paroist dans le Zodiaque*. A Paris, De L'Imprimerie Royale.

Chambers, R. (1846). *Explanations: A Sequel to "Vestiges of the Natural History of Creation"* (Wiley & Putnam).

Chesley, S.R. et al. (2014). Orbit and bulk density of the OSIRIS-REx target Asteroid (101955) Bennu. *Icarus* 235, 5–22.

Chesley, S. R. et al. (2020). Trajectory Estimation for Particles Observed in the Vicinity of (101955) Bennu. *J. Geophys. Res. Planet.* 125, e2019JE006363.

Childrey, J. (1661). *Britania Baconica: or, the natural rarities of England, Scotland, & Wales* (self-published).

Chladni, E.F.F. (1794). *Ueber den Ursprung der von Pallas gefundenen und anderer ihr* ähnlicher *Eisenmassen und* über *einige damit in Verbindung stehende Naturerscheinungen* (Bey Johann Friedrich Hartknoch).

Christensen, P.R. et al. (2018). The OSIRIS-REx Thermal Emission Spectrometer (OTES) Instrument. *Space Sci. Rev.* 214, 87.

Clark, B.E. et al. (2011). Asteroid (101955) 1999 RQ36: Spectroscopy from 0.4 to 2.4 μm and meteorite analogs. *Icarus* 216, 462–475.

Connelly, J.N. et al. (2017). Pb–Pb chronometry and the early Solar System. *Geochim. Cosmochim. Acta* 201, 345–363.

Coradini, M. (2001). Planetary exploration. In *Encyclopedia of Astronomy & Astrophysics*, Murdin, P., Ed. (CRC Press).

Daly, M.G. et al. (2017). The OSIRIS-REx Laser Altimeter (OLA) Investigation and Instrument. *Space Sci. Rev.* 212, 899–924.

Daly, M.G. et al. (2020). Hemispherical differences in the shape and topography of asteroid (101955) Bennu. *Sci. Adv.* 6, eabd3649.

Daly, R.T. et al. (2020). The morphometry of impact craters on Bennu. *Geophys. Res. Lett.* 47, e2020GL089672.

DellaGiustina, D.N. et al. (2018). Overcoming the Challenges Associated with Image-Based Mapping of Small Bodies in Preparation for the OSIRIS-REx Mission to (101955) Bennu. *Earth Space Sci.* 5, 929–949.

DellaGiustina, D.N., Emery, J.P. et al. (2019). Properties of rubble-pile asteroid (101955) Bennu from OSIRIS-REx imaging and thermal analysis. *Nat. Astron.* 3, 341–351.

DellaGiustina, D.N. et al. (2020). Variations in color and reflectance on the surface of asteroid (101955) Bennu. *Science* 370, eabc3660.

DellaGiustina, D.N., Kaplan, H.H. et al. (2021). Exogenic basalt on asteroid (101955) Bennu. *Nat. Astron.* 5, 31–38.

Delbo, M. et al. (2022). Alignment of fractures on Bennu's boulders indicative of rapid asteroid surface evolution. *Nat. Geosci.* 15, 453–457.

DeMeo, F.E. & Carry, B. (2014). Solar System evolution from compositional mapping of the asteroid belt. *Nature* 505, 629–634.

Deshapriya, J.D.P. et al. (2021). Spectral Analysis of Craters on (101955) Bennu. *Icarus* 357, 114252.

Duxbury, T.C. et al. (2004). Asteroid 5535 Annefrank size, shape, and orientation: Stardust first results. *J. Geophys. Res. Planet.* 109, E02002.

Emery, J.P. et al. (2014). Thermal infrared observations and thermophysical characterization of OSIRIS-REx target asteroid (101955) Bennu. *Icarus* 234, 17–35.

Farnocchia, D. et al. (2021). Ephemeris and hazard assessment for near-Earth asteroid (101955) Bennu based on OSIRIS-REx data. *Icarus* 369, 114594.

Farrington, O.C. (1900). The worship and folk-lore of meteorites. *J. Am. Folklore* 13, 199–208.

Fornasier, S. et al. (2020). Phase Reddening on Asteroid Bennu from Visible and Near-Infrared Spectroscopy. *Astron. Astrophys.* 644, A142.

Fujiwara, A. et al. (2006). The rubble-pile asteroid Itokawa as observed by Hayabusa. *Science* 312, 1330–1334.

Gehrels, T., Ed. (1979). *Asteroids* (Univ. Arizona Press).

Golish, D.R. et al. (2020). Ground and In-Flight Calibration of the OSIRIS-REx Camera Suite. *Space Sci. Rev.* 216, 12.

Golish, D.R. et al. (2021). Disk-Resolved Photometric Modeling and Properties of Asteroid (101955) Bennu. *Icarus* 357, 113724.

Golish, D.R. et al. (2021). A high-resolution normal albedo map of asteroid (101955) Bennu. *Icarus* 355, 114133.

Gulkis, S. et al. (2010). Millimeter and submillimeter measurements of asteroid (2867) Steins during the Rosetta fly-by. *Planet. Space Sci.* 58, 1077–1087.

Hamilton, V.E. et al. (2019). Evidence for widespread hydrated minerals on asteroid (101955) Bennu. *Nat. Astron.* 3, 332–340.

Hamilton, V.E. et al. (2021). Evidence for limited compositional and particle size variation on asteroid (101955) Bennu from thermal infrared spectroscopy. *Astron. Astrophys.* 650, A120.

Hencke, M. (1846). Announcement of his discovery of Astréa. *Mon. Not. R. Astron. Soc.* 7, 27.

Hergenrother, C.W. et al. (2019). The operational environment and rotational acceleration of asteroid (101955) Bennu from OSIRIS-REx observations. *Nat. Commun.* 10, 1291.

Hergenrother, C.W. et al. (2020). Introduction to the Special Issue: Exploration of the Activity of Asteroid (101955) Bennu. *J. Geophys. Res. Planet.* 125, e2020JE006549.

Hergenrother, C.W. et al. (2020). Photometry of Particles Ejected from Active Asteroid (101955) Bennu. *J. Geophys. Res. Planet.* 125, e2020JE006381.

Herschel, W. (1802). VIII. Observations on the two lately discovered celestial bodies. *Phil. Trans. R. Soc. London* 92, 213–232.

Hoyle, F. (1960). The origin of the solar nebula. *Quart. J. R. Astron. Soc.* 1, 28.

Jambon, A. (2017). Bronze Age iron: Meteoritic or not? A chemical strategy. *J. Archaeol. Sci.* 88, 47–53.

Jawin, E.R. et al. (2020). Global Patterns of Recent Mass Movement on Asteroid (101955) Bennu. *J. Geophys. Res. Planet.* 125, e2020JE006475.

Jawin, E.R. et al. (2022). Global geologic map of asteroid (101955) Bennu indicates heterogeneous resurfacing in the past 500,000 years. *Icarus* 381, 114992.

Ji, J. et al. (2015). Chang'e-2 spacecraft observations of asteroid 4179 Toutatis. *Proc. Intl. Astron. Union* 10, 144–152.

Johnson, D. et al. (2013). Analysis of a prehistoric Egyptian iron bead with implications for the use and perception of meteorite iron in ancient Egypt. *Meteorit. Planet. Sci.* 48, 997–1006.

Kaplan, H.H. et al. (2020). Bright carbonate veins on asteroid (101955) Bennu: Implications for aqueous alteration history. *Science* 370, eabc3557.

Kaplan, H.H. et al. (2021). Composition of organics on asteroid (101955) Bennu. *Astron. Astrophys.* 653, L1.

King, E.A. et al. (1969). Meteorite fall at Pueblito de Allende, Chihuahua, Mexico: Preliminary information. *Science* 163, 928–929.

Kruijer, T.S. et al. (2020). The great isotopic dichotomy of the early Solar System. *Nat. Astron.* 4, 32–40.

Kuiper, G.P. (1950). On the origin of asteroids. *Astron. J.* 55,164.

Kuiper, G.P. (1951). On the origin of the solar system. *Proc. Natl. Acad. Sci. U.S.A.* 37, 1.

Kvenvolden, K. et al. (1970). Evidence for extraterrestrial amino-acids and hydrocarbons in the Murchison meteorite. *Nature* 228, 923–926.

Lauretta, D.S. & Kilgore, M. (2005). *A Color Atlas of Meteorites in Thin Section.* (Golden Retriever Publications).

Lauretta, D.S. et al. (2015). The OSIRIS-REx target asteroid (101955) Bennu: Constraints on its physical, geological, and dynamical nature from astronomical observations. *Meteorit. Planet. Sci.* 50, 834–849.

Lauretta, D.S. et al. (2017). OSIRIS-REx: Sample Return from Asteroid (101955) Bennu. *Space Sci. Rev.* 212, 925–984.

Lauretta, D.S., DellaGiustina, D.N. et al. (2019). The unexpected surface of asteroid (101955) Bennu. *Nature* 568, 55–60.

Lauretta, D.S., Hergenrother, C.W. et al. (2019). Episodes of particle ejection from the active asteroid (101955) Bennu. *Science* 366, eaay3544.

Lauretta, D.S. et al. (2021). OSIRIS-REx at Bennu: Overcoming challenges to collect a sample of the early Solar System. In *Sample Return Missions*, Longobardo, A., Ed. (Elsevier), pp. 163–194.

Lauretta, D.S. et al. (2022). Spacecraft sample collection and subsurface excavation of asteroid (101955) Bennu. *Science* 377, 285–291.

Le Corre, L. et al. (2021). Characterization of Exogenic Boulders on the Near-Earth Asteroid (101955) Bennu from OSIRIS-REx Color Images. *Planet. Sci. J.* 2, 114.

Leonard, F.C. (1949). Some remarks on the origin of earthly meteorites. *Contrib. Meteorit. Soc.* 4, 212–214.

Li, J.-Y. et al. (2021). Spectrophotometric Modeling and Mapping of (101955) Bennu. *Planet. Sci. J.* 2, 117.

Lord III, H. C. (1965). Molecular equilibria and condensation in a solar nebula and cool stellar atmospheres. *Icarus* 4, 279–288.

Lunar Sample Preliminary Examination Team (1969). Preliminary examination of lunar samples from Apollo 11. *Science* 165, 1211–1227.

Marvin, U.B. (1983). The discovery and initial characterization of Allan Hills 81005: The first lunar meteorite. *Geophys. Res. Lett.* 10, 775–778.

Marvin, U.B. (1996). Ernst Florens Friedrich Chladni (1756–1827) and the origins of modern meteorite research. *Meteorit. Planet. Sci.* 31, 545–588.

Masterson, R.A. et al. (2018). Regolith X-Ray Imaging Spectrometer (REXIS) Aboard the OSIRIS-REx Asteroid Sample Return Mission. *Space Sci. Rev.* 214, 48.

Mattinson, J.M. (2013). Revolution and evolution: 100 years of U–Pb geochronology. *Elements* 9, 53–57.

May, B. H. (2007). *A Survey of Radial Velocities in the Zodiacal Dust Cloud.* PhD thesis, Imperial College of Science, Technology and Medicine, London.

McBeath, A. & Gheorghe, A.D. (2005). Meteor beliefs project: Meteorite worship in the ancient Greek and Roman worlds. *WGN J. Intl. Meteor Org.* 33, 135–144.

McCoy, T.J. et al. (2017). The Anoka, Minnesota iron meteorite as parent to Hopewell meteoritic metal beads from Havana, Illinois. *J. Archaeol. Sci.* 81, 13–22.

Meltzer, M. (2015). The interplanetary journey. In *The Cassini-Huygens Visit to Saturn: An Historic Mission to the Ringed Planet* (Springer), pp. 181–212.

Michel, P., Ballouz, R.-L. et al. (2020). Collisional formation of top-shaped asteroids and implications for the origins of Ryugu and Bennu. *Nat. Commun.* 11, 2655.

Miyamoto, M. (1991). Thermal metamorphism of CI and CM carbonaceous chondrites: An internal heating model. *Meteoritics* 26, 111–115.

Mojsov, B. (2005). *Osiris: Death and Afterlife of a God* (Wiley-Blackwell).

Molaro, J.L. et al. (2020). In situ evidence of thermally induced rock breakdown widespread on Bennu's surface. *Nat. Commun.* 11, 2913.

Molaro, J.L. et al. (2020). Thermal Fatigue as a Driving Mechanism for Activity on Asteroid Bennu. *J. Geophys. Res. Planet.* 125, e2019JE006325.

Nolan, M.C. et al. (2013). Shape model and surface properties of the OSIRIS-REx target Asteroid (101955) Bennu from radar and lightcurve observations. *Icarus* 226, 629–640.

Nolan, M.C. et al. (2019). Detection of Rotational Acceleration of Bennu Using HST Light Curve Observations. *Geophys. Res. Lett.* 46, 1956–1962.

Norman, C.D. et al. (2022). Autonomous Navigation Performance Using Natural Feature Tracking during the OSIRIS-REx Touch-and-Go Sample Collection Event. *Planet. Sci. J.* 3, 101.

Oberst, J. et al. (2001). A model for rotation and shape of asteroid 9969 Braille from ground-based observations and images obtained during the Deep Space 1 (DS1) flyby. *Icarus* 153, 16–23.

Olds, R. et al. (2022). The Use of Digital Terrain Models for Natural Feature Tracking at Asteroid Bennu. *Planet. Sci. J.* 3, 100.

Palmer, E. E. et al. (2022). Practical Stereophotoclinometry for Modeling Shape and Topography on Planetary Missions. *Planet. Sci. J.* 3, 102.

Perry, M. E. et al. (2022). Low surface strength of the asteroid Bennu inferred from impact ejecta deposit. *Nat. Geosci.* 15, 447–452.

Planetary Society (2013). Nine-Year-Old Names Asteroid Target of NASA Mission in Competition Run By The Planetary Society (press release); planetary.org/press-room/releases/2013/nine-year-old-names-asteroid.html.

Praet, A. et al. (2021). Hydrogen abundance estimation and distribution on (101955) Bennu. *Icarus* 363, 114427.

Prockter, L. et al. (2002). The NEAR Shoemaker mission to asteroid 433 Eros. *Acta Astronaut.* 51, 491–500.

Raymond, S.N., & Izidoro, A. (2017). Origin of water in the inner Solar System: Planetesimals scattered inward during Jupiter and Saturn's rapid gas accretion. *Icarus* 297, 134–148.

Reuter, D.C. et al. (2018). The OSIRIS-REx Visible and InfraRed Spectrometer (OVIRS): Spectral Maps of the Asteroid Bennu. *Space Sci. Rev.* 214, 54.

Rizk, B. et al. (2018). OCAMS: The OSIRIS-REx Camera Suite. *Space Sci. Rev.* 214, 26.

Robertson, H.P. (1937). Dynamical effects of radiation in the solar system. *Mon. Not. R. Astron. Soc.* 97, 423.

Rozitis, B. et al. (2020). Implications for ice stability and particle ejection from high-resolution temperature modeling of asteroid (101955) Bennu. *J. Geophys. Res. Planet.* 125, e2019JE006323.

Rozitis, B. et al. (2020). Asteroid (101955) Bennu's weak boulders and thermally anomalous equator. *Sci. Adv.* 6, eabc3699.

Rozitis, B. et al. (2022). High-Resolution Thermophysical Analysis of the OSIRIS-REx Sample Site and Three Other Regions of Interest on Bennu. *J. Geophys. Res. Planet.* 127, e2021JE007153.

Russell, C., & Raymond, C. (2012). *The Dawn Mission to Minor Planets 4 Vesta and 1 Ceres* (Springer).

Scheeres, D.J. et al. (2019). The dynamic geophysical environment of (101955) Bennu based on OSIRIS-REx measurements. *Nat. Astron.* 3, 352–361.

Scheeres, D.J. et al. (2020). Heterogenous mass distribution of the rubble-pile asteroid (101955) Bennu. *Sci. Adv.* 6, eabc3350.

Schröter, J.H. (1805). *Lilienthalische Beobachtungen der neu entdeckten Planeten Ceres, Pallas und Juno, zur genauen und richtigen Kenntniss ihrer wahren Grössen, Atmosphären und übrigen merkwürdigen Naturverhältnisse im Sonnengebiete.*

Schröter, J.H. (1807). XI. Observations and measurements of the planet Vesta. *Phil. Trans. R. Soc. London* 97, 245–246.

Schultz, R. et al. (2012). Rosetta fly-by at asteroid (21) Lutetia: An overview. *Planet. Space Sci.* 66, 2–8.

Seargent, D.A.J. (1990). The Murchison meteorite: Circumstances of its fall. *Meteoritics* 25, 341–342.

Shiraishi, K. (1979). *Antarctic Search for Meteorite by U.S.-Japan Joint Party*, 1978–1979. National Institute for Polar Research, Tokyo, Japan.

Simon, A. A. et al. (2020). Widespread carbon-bearing materials on near-Earth asteroid (101955) Bennu. *Science* 370, eabc3522.

Simon, A. A. et al. (2020). Weak spectral features on (101995) Bennu from the OSIRIS-REx Visible and InfraRed Spectrometer. *Astron. Astrophys.* 644, A148.

Simon, A.A. et al. (2021). Derivation of the final OSIRIS-REx OVIRS in-flight radiometric calibration. *JATIS* 7, 020501.

Sorby, H.C. (1877). On the structure and origin of meteorites. *Nature* 15, 495–498.

Steele, R C.J. et al. (2017). Matrix effects on the relative sensitivity factors for manganese and chromium during ion microprobe analysis of carbonate: Implications for early Solar System chronology. *Geochim. Cosmochim. Acta* 201, 245–259.

Tatsumi, E. et al. (2021). Widely Distributed Exogenic Materials of Varying Compositions and Morphologies on Asteroid (101955) Bennu. *Mon. Not. R. Astron. Soc.* 508, 2053–2070.

Tholen, D.J. (1989). Asteroid taxonomic classifications. In *Asteroids II*, Binzel, R.P. et al., Eds. (Univ. Arizona Press), pp. 1139–1150.

Tricarico, P. et al. (2021). Internal Rubble Properties of Asteroid (101955) Bennu. *Icarus* 370, 14665.

Veverka, J. et al. (1994). Galileo's encounter with 951 Gaspra: Overview. *Icarus* 107, 2–17.

Waldrop, M. M. (1982). Operation spacewatch. *Science* 216, 42.

Walsh, K. J. et al. (2013). Introducing the Eulalia and new Polana asteroid families: Re-assessing primitive asteroid families in the inner Main Belt. *Icarus* 225, 283–297.

Walsh, K.J. (2018). Rubble Pile Asteroids. *Annu. Rev. Astron. Astrophys.* 56, 593–624.

Walsh, K.J. et al. (2019). Craters, boulders and regolith of (101955) Bennu indicative of an old and dynamic surface. *Nat. Geosci.* 12, 242–246.

Walsh, K.J. et al. (2022). Assessing the Sampleability of Bennu's Surface for the OSIRIS-REx Asteroid Sample Return Mission. *Space. Sci. Rev.* 218, 20.

Walsh, K.J. et al. (2022). Near-zero cohesion and loose packing of Bennu's near subsurface revealed by spacecraft contact. *Sci. Adv.* 8, eabm6229.

Warren, P.H. (2011). Stable-isotopic anomalies and the accretionary assemblage of the Earth and Mars: A subordinate role for carbonaceous chondrites. *Earth Planet. Sci. Lett.* 311, 93–100.

Watanabe, S. et al. (2017). Hayabusa2 mission overview. *Space Sci. Rev.* 208, 3–16.

Watanabe, S. et al. (2019). Hayabusa2 arrives at the carbonaceous asteroid 162173 Ryugu-A spinning top-shaped rubble pile. *Science* 364, 268–272.

Weisberg, M.K. et al. (2006). Systematics and evaluation of meteorite classification. In *Meteorites and the Early Solar System II*, Lauretta, D.S. et al., Eds. (Univ. Arizona Press), pp. 13–52.

Wilkinson, R.H. (2003). *The Complete Gods and Goddesses of Ancient Egypt* (Thames and Hudson).

Williams, G. (2001). Asteroid Discovery in History. In *Encyclopedia of Astronomy and Astrophysics*, Murdin, P., Ed. (CRC Press), p. 2747.

Williams, B. et al. (2018). OSIRIS-REx Flight Dynamics and Navigation Design. *Space Sci. Rev.* 214, 69.

Yoshida, M. (1971). Discovery of meteorites near Yamato mountains, east Antarctica. *Polar Sci.* 3, 62–65.

Zou, X.D. et al. (2021). Photometry of Asteroid (101955) Bennu with OVIRS on OSIRIS-REx. *Icarus* 358, 114183.

Online references

Gazetteer of Planetary Nomenclature: planetarynames.wr.usgs.gov
Impact Calculator: eaps.purdue.edu/impactcrater
Meteoritical Bulletin Database: lpi.usra.edu/meteor/metbull.php
Minor Planet Center: minorplanetcenter.net
NASA Scientific Visualization Studio: svs.gsfc.nasa.gov
NASA Sentry Earth Impact Monitoring: cneos.jpl.nasa.gov/sentry
OSIRIS-REx mission: asteroidmission.org

Additional works consulted for the Glossary

Coles, K.S. et al. (2019). *Atlas of Mars: Mapping Its Geography and Geology* (Cambridge Univ. Press).

Lauretta, D.S., & McSween, H.Y., Eds. (2006). *Meteorites and the Early Solar System II* (Univ. Arizona Press).

McFadden, L. et al. (2007). *Encyclopedia of the Solar System* (Elsevier, ed. 2).

Moore, P. (2002). *Astronomy Encyclopedia* (Oxford Univ. Press).

Neuendorf, K.K.E. et al., Eds. (2011). *Glossary of Geology* (American Geological Institute, ed. 5).

Glossary

accretion: a process in which small particles or bodies come together and gradually accumulate to form a larger body. In planetary science, the growth of celestial bodies such as asteroids, planets, and natural satellites through the gradual accumulation of matter via gravitational attraction.

achondrite: A type of stony *meteorite* that formed by melting and recrystallization within an asteroid. They have distinct textures and mineralogies indicative of igneous processes and are also known as differentiated stony meteorites.

albedo: The ratio of the electromagnetic radiation reflected by a body to the amount incident upon it, often expressed as a percentage.

aphelion: The point in a body's elliptical orbit at which it is farthest from the Sun.

Apollo-type: A *near-Earth object* with an Earth-crossing orbit having a *semi-major axis* larger than Earth's.

asteroid (also **minor planet**): A class of small natural bodies that follow independent orbits around the Sun and which appear in telescopes as unresolved *point sources* when present in the inner Solar System.

asteroid family: Asteroids grouped by similarity in their orbits; believed to originate from an asteroid collision that produced *rubble piles* and individual fragments.

astronomical unit (AU): Average distance between Earth and the Sun; unit of measurement for Solar System distances.

baseline: Distance between the camera positions in the two images composing a stereo pair.

Bode's Law (also **Titius-Bode Law**): Numerical relationship matching the distances of planets known in 1772 from the Sun. A planet was missing in the sequence at 2.8 AU, prompting observations that led to the discovery of the *main asteroid belt*.

brecciation: The process of breaking rocks into fragments, which are then compacted, cemented, or recrystallized to form a new, amalgamated rock. The resulting rock is a breccia.

burn: Spacecraft maneuver using the propulsion system to change direction or velocity.

carbonaceous: Made of, containing, or related to carbon. This term is commonly used to describe materials that are rich in organic carbon, such as coal, oil, and peat. In a broader sense, it can describe any material or substance that contains carbon, including living organisms and many minerals.

carbonate: A mineral having carbonate ion (CO_3) in its structure; a rock formed of or containing carbonate minerals (some terrestrial examples include limestone, marble, and travertine).

chondrite (adjective **chondritic**): Originally defined as a *meteorite* that contained *chondrules*; now also implies a bulk chemical composition, for all but the most volatile elements, that is not far removed from that of the Sun.

chondrule: Approximately spherical assemblages, characteristic of most *chondrites*, that existed independently prior to incorporation in the *meteorite* and that show evidence for partial or complete melting.

counts (as used in map legends in Chapter 7): The number of facets on the global shape model of Bennu that have the given value; a measure of relative areal coverage. Different versions of the shape model were used depending on the data mapped, so total counts are not necessarily identical between maps. For the gravity anomaly map, "counts" refers to the number of grid squares in a two-dimensional gridded map.

crater: A hole or depression. Most are roughly circular or oval in outline. On Bennu, craters are likely of impact origin.

DEC (declination): Measure of angular distance north or south of the celestial equator (projection of Earth's equator onto the sky); one of the two coordinates of the celestial coordinate system, the other being *right ascension*.

digital terrain model: A coordinate grid of values representing height or elevation. See also *shape model*.

Doppler tracking: Velocity determination in which the frequency shift of the Doppler effect is directly related to the relative motion between the transmitter and receiver.

dynamical resonance: The characteristic of two objects orbiting a central body whose periods have a small integer ratio; for example, Jupiter and an asteroid orbiting the Sun. Dynamical resonance can create nonlinear effects that push the smaller object out of its orbit.

eccentricity: A measure of the elongation of an ellipse, zero for a perfect circle and approaching 1 as the elongation increases. Used to describe the shape of orbits.

ecliptic: The mean plane of Earth's orbit around the Sun, used as the reference plane from which the *inclinations* of other planetary orbits are defined.

exogenic (also **exogenous**): Originating from a different planetary body or *asteroid family* than where it is found.

facet: The smallest resolved unit of surface area making up a three-dimensional *shape model*; one face of a polyhedron. Global shape models of Bennu consist of triangular facets with vertex separations of less than 1 meter.

geopotential: A surface on which the *gravitational potential* is constant at all locations.

GM: The universal gravitational constant (G) multiplied by the planetary mass (M). This parameter is used commonly in planetary science as a convenient way to indicate the strength of an object's gravitational field.

gravity anomaly (also **Bouguer anomaly**): Difference between measured gravity and that expected after corrections are made for elevation and density differences from the *geopotential*.

gravitational potential: The potential energy of a unit mass at a particular point in a field of gravitational attraction. Differentiating the potential gives the acceleration; mass tends to accelerate toward lower (more negative) potential.

Hertz (Hz): The standard international unit of frequency; 1 Hertz = 1 event, cycle, or wave per second.

Hill sphere: A sphere with a radius equal to the distance from the body at which the gravitational influence of another celestial body becomes stronger than that of the body itself. The size of a planet's Hill sphere depends on its mass, its distance from the Sun, and the masses of other nearby celestial bodies.

hydroxyl: A combination of an oxygen atom and a hydrogen atom (OH-); if negatively charged, it is an ion.

impact shock: The shock wave, or its effects, arising from the collision of bodies at or near orbital velocities.

inclination: The angle between the plane of an orbit and a reference plane, here usually the *ecliptic*.

invariable plane: Plane of reference that is at right angles to the total angular momentum vector of the Solar System; it is inclined at a few degrees from the *ecliptic*.

lidar: Light detection and ranging; measurement of distance to an object by the time between transmission and reception of a reflected light signal (commonly laser).

magnetite: A naturally occurring iron oxide mineral with strong magnetic properties and the chemical formula Fe_3O_4. Found in a variety of geological settings, including igneous, metamorphic, and hydrothermally altered rocks.

main asteroid belt (also **main belt** or **asteroid belt**): A region between the orbits of Mars and Jupiter occupied by numerous *asteroids* in dynamically stable orbits.

map projection: A method to portray a part of a spherical or irregular body on a flat surface (paper or digital). A flat map can show one or more, but not all, of the following: true directions, true distances, true areas, and/or true shapes.

mass movement (also **mass flow**): The downslope movement of rock or *regolith* material, solely under the influence of gravity; includes slow displacement such as creep and rapid displacements such as slides and avalanches.

meteorite: A natural object of extraterrestrial origin that survives passage through a planetary atmosphere.

meteoroid: A natural small object in an independent orbit in the Solar System.

near-Earth asteroid (NEA): An *asteroid* in an orbit that approaches Earth's, specifically with *perihelion* less than 1.3 AU.

near-Earth object (NEO): Asteroid or comet in an orbit that approaches Earth's, officially with *perihelion* less than 1.3 AU. In practice, denotes objects with smaller perihelia making very close approaches to Earth feasible.

non-principal axis rotation: Having rotation that is not around a single, principal axis. An *asteroid* with non-principal axis rotation can be tumbling, complicating efforts to approach it for sampling.

obliquity: The angle of a body's rotational axis relative to the plane of its orbit. An axis of a body in *prograde* rotation at a right angle to the plane of orbit has an obliquity of 0°.

parent body (also **parent asteroid**): The coherent object from which a given *meteorite,* class of meteorites, *asteroid family*, or *rubble pile* originated.

particle size frequency distribution: The percentage, by mass or by number, of particles in each mass or size fraction of a sample or population.

perihelion: The point in a body's elliptical orbit at which it is nearest to the Sun.

phase angle: Angle between the Sun, a given object, and the observer, with the object at the vertex.

phase reddening: An effect that produces an increase of the *spectral slope* and variations in the strength of the absorption bands as the *phase angle* increases. The effect is explained as the wavelength dependence of scattering of light.

phyllosilicate: Minerals in which the tetrahedral silicate (SiO_4) units are arranged in flat sheets; includes clay minerals.

pixel: In a digital image, the area represented by a single digital value. See also *resolution*.

planetesimal: A solid celestial body that accumulated during the first stages of planet formation. Planetesimals aggregated into progressively larger bodies to form planets.

planetocentric: Denotes latitude that is measured as the angle between the equatorial plane and a vector directed at a point of interest, measured at the center of mass of the planetary body.

point source: A celestial object that is so distant and/or small that it appears as a single, unresolved point of light when observed with a telescope.

precession: Gradual motion of the axis of rotation of a planetary body that causes the position of the celestial poles to describe a circle. For Earth, the period of circular motion is about 26,000 years and is largely caused by the gravitational pull of the Sun and Moon on Earth's equatorial bulge. For smaller bodies, the motion of the axis of rotation can be more complex.

prograde: Rotation or revolution of a body in the same direction as the dominant angular momentum of the Solar System. The direction is counterclockwise as viewed from above the north pole of the Sun.

protoplanetary disk: A dense, rotating disk of gas, ice, and dust that surrounds a young star and is the birthplace of planets. Protoplanetary disks are thought to be the result of the collapse of a molecular cloud. The disk material eventually begins to clump and form *planetesimals*, which then grow into full-fledged planets through the process of *accretion*. See also *solar nebula*.

protostar: A dense, gravitationally collapsing cloud of gas and dust in an early stage of star formation. As a protostar collapses, its temperature and density increase until the core becomes hot enough to ignite hydrogen fusion, at which point it becomes a star.

pyroxene: Silicate mineral that is common in igneous rocks, having calcium, sodium, magnesium, iron, or other cations with silicon and oxygen.

RA (right ascension): Measure of angular distance along the celestial equator (projection of Earth's equator onto the sky); one of the two coordinates of the celestial coordinate system, the other being *declination*.

regio (plural **regiones**): From Latin; formal term for a broad geographic region, used in feature names.

regolith: Layer of loose, fragmented material that covers solid rock on the surface of celestial bodies, such as planets, moons, and asteroids. Regolith is formed by processes such as weathering, erosion, and impact events, and is composed of a mixture of rocks, minerals, and dust.

resolution: A measure of the ability of a remote-sensing system to distinguish detail or define closely spaced targets; the minimum size of a feature that can be detected. See also *pixel*.

retrograde: Rotation or orbital motion in the opposite direction of *prograde* motion.

rubble pile: A collection of otherwise loose fragments held together by gravity. Rubble-pile *asteroids* typically have higher porosity, and thus lower density, than entirely solid bodies.

saxum (plural **saxa**): From Latin; formal term for a boulder or rock, used in feature names.

semi-major axis: The mean separation between two bodies in an elliptical orbit; half the length of the major axis of an ellipse.

shape model: A *digital terrain model* of a celestial body, composed of *facets*.

sidereal period: The period measured relative to the stars, as for the rotation or revolution of a body.

solar nebula: The *protoplanetary disk* from which the Solar System formed.

spectrum: Visible light or other electromagnetic radiation, arrayed according to its wavelengths (colors).

spectral signature (also **spectral fingerprint**): A distinctive or diagnostic pattern of reflected (light) radiation as a function of wavelength. Different materials whose spectral signatures contrast can be identified or distinguished using reflected light.

spectral slope: The ratio of the change in intensity of light as a function of its wavelength. A positive spectral slope indicates that red (or the long wavelength measured) is stronger than the other, shorter-wavelength colors in the spectrum; a negative slope occurs when blue (or the short wavelength) is stronger.

standard deviation: A measure of the amount of variation in a set of numeric values. A low standard deviation indicates that individual values tend to be close to the average of the values.

surface roughness: A measure of the deviations of a surface from an ideal or smooth form.

synodic period: The interval between successive alignments of a body with Earth and the Sun.

thermal inertia: Measure of the rate of response of a material to temperature changes. A material that heats or cools very slowly has high thermal inertia.

Tholen classification: A system of letter designation to distinguish *asteroids* having distinct, characteristic colors, originally proposed by Tholen in 1989 and modified by others.

Yarkovsky effect: A non-gravitational thrust produced when small bodies absorb sunlight, heat up, and then re-radiate the energy after a short delay caused by *thermal inertia*. Produces a slow but steady drift in the *semi-major axis* of an object's orbit.

YORP (**Yarkovsky-O'Keefe-Radzievskii-Paddack**) **effect**: A torque created by asymmetry in the absorption and reemission of solar energy by a rotating body. The YORP effect can change the rotation rate and drive the rotation pole to a position perpendicular to the sunlight.

Author biographies

Dante Lauretta is a Regents Professor in the Lunar and Planetary Laboratory at the University of Arizona in Tucson, AZ. He holds a B.S. in Mathematics and Physics and a B.A. in East Asian Studies from the University of Arizona. He completed his Ph.D. in Earth and Planetary Sciences at Washington University in St. Louis. His research interests include the formation and evolution of the Solar System and the origin of life, with an emphasis on laboratory studies of meteorites and returned extraterrestrial samples, as well as spacecraft exploration of asteroids and comets.

Lauretta is Principal Investigator for NASA's OSIRIS-REx asteroid sample return mission. He is also a Co-Investigator on JAXA's Hayabusa2 asteroid sample return mission and ESA's Hera mission to asteroid Didymos. His research uses astronomical observations, spacecraft-based remote sensing, and laboratory measurements of extraterrestrial materials.

Lauretta received the University of Arizona Galileo Circle Dean's Award in 2020. He was elected a Fellow of the Meteoritical Society in 2014. He was awarded the NASA Silver Achievement Medal in 2017 for his role on the OSIRIS-REx Asteroid Astronomy team. In 2017 he was also inducted into Boys and Girls Clubs of America Alumni Hall of Fame.

Sir Brian May, CBE, Ph.D, ARCS, FRAS is a founding member of the rock group Queen, a world-renowned guitarist, songwriter, producer and performer, 3-D stereoscopic photographic authority, author, publisher, and passionate campaigner for animal rights. On graduating from Imperial College London in 1968 with a B.Sc. (Hons) degree in physics, Brian began a Ph.D. in Astrophysics, recording high-resolution spectra of the Zodiacal Light using a home-spun Fabry-Perot Spectrometer. In 1974, when his musical career with Queen took over, Brian was forced to shelve his Ph.D. work, but in 2006, with the encouragement of Professor Michael Rowan Robinson, Professor Francisco Sanchez Martinez, Dr Garik Israelian, and Sir Patrick Moore, he returned to complete his studies.

In 2007 Brian was awarded his full Ph.D. degree in Astrophysics for his thesis: *A Survey of Radial Velocities in the Zodiacal Dust Cloud*. Thirty-seven years had elapsed between registration and submission of the Ph.D. He currently holds the post of visiting researcher at Imperial to continue his work in astronomy. He is also a cofounder of the planetary defence awareness campaign, Asteroid Day. In 2008 Asteroid 52665 Brianmay was named after him. His thesis was published by Springer Verlag in 2007, and he has since written, edited, or published 14 other books on astronomy, Queen, and stereo photography.

In 2015, Brian was appointed Science Team Collaborator by the *New Horizons'* Principal Investigator Alan Stern, co-producing the world's first stereo images of Pluto and Kuiper Belt Object Arrokoth. To celebrate the latter encounter, Brian composed and released a single called *New Horizons* on New Year's day 2019. In 2018 he worked with the Rosetta mission's Matt Taylor and Joel Parker to create, along with his astro-collaborator Claudia Manzoni, the first stereo images of comet 67P/Churyumov-Gerasimenko. In Spring 2019, working with University of Arizona Professor Dante Lauretta using the data from NASA's *OSIRIS-REx* mission, Brian and Claudia created the first stereo images of the asteroid Bennu and were adopted as Science Team Collaborators for this mission.

Carina A. Bennett is a Project Manager and Software Engineer in the Lunar and Planetary Laboratory at the University of Arizona. She holds a B.A. in Media Arts and Creative Writing, a B.S. in Computer Science, and a M.S. in Computer Science from the University of Arizona as well as a M.F.A in Film Production from the University of Iowa. She is a collaborator on the OSIRIS-REx mission to Bennu and previously worked as a Senior Image Processing Engineer as part of the mission's Image Processing Working Group, where she developed software to automate the production of controlled mosaics and thematic maps for the mission. She has extensive experience working with images of irregularly shaped small planetary bodies and has advised the US Geological Survey's Astrogeology Science Center on the capabilities required to successfully implement 3-D shape model support. In 2019 Bennett was awarded the OSIRIS-REx PI Award of Distinction and has been nominated for a Rocky Mountain Regional Emmy for her video production work.

Kenneth S. Coles is Professor in the Department of Geography, Geology, Environment, and Planning and Planetarium Director at Indiana University of Pennsylvania, where he teaches astronomy and planetary geology courses. He holds a B.S. and M.S. in geology from Caltech and a Ph.D. in geology from Columbia University. For thirty years he has specialized in presenting planetary science discoveries to undergraduate students, schoolchildren, and the general public.

Among the projects he has conducted are coordinated observations by schoolchildren of solar eclipses, promoting measurements of asteroid occultations by amateur astronomers, and giving free planetarium programs to thousands of adults and children on a variety of astronomical topics. His educational outreach and curriculum development has been funded by grants from the National Science Foundation and NASA. He is the lead author of *The Atlas of Mars: Mapping Its Geography and Geology* (Cambridge University Press, 2019), which won the 2020 Prose Award of the Association of American Publishers for Cosmology and Astronomy.

Indiana University of Pennsylvania has recognized Coles with awards for Excellence in Expository Instruction, Career Service for public outreach in astronomy, and as Outstanding Researcher in science.

Claudia Manzoni is an amateur astro-stereographer who enjoys composing stereo views of Solar System objects from images in space agencies' archives — something she has been collaborating on with Sir Brian May for almost ten years. This passion led her to contribute in the role of 3-D compositor and researcher to *Mission Moon 3-D: Reliving the Great Space Race* by David J. Eicher and Brian May, published by the London Stereoscopic Company and MIT Press in October 2018.

Along with Sir Brian, she is a collaborator on the OSIRIS-REx mission to Bennu, the Hayabusa2 mission to Ryugu, and the Mastcam-Z team of the Mars 2020 Perseverance Rover mission.

She holds an M.Sc. in Chemical Engineering from the University of Bologna and is a material researcher in a chemical company.

C. W. V. Wolner, also of the Lunar and Planetary Laboratory at the University of Arizona, is the Chief Editor of the OSIRIS-REx mission, overseeing the scientific publications produced by the team and writing about their findings for different audiences. Wolner completed a B.A. at Oberlin College and the University of Aberdeen, and a M.S. at the University of Virginia, both in Earth sciences. She has also taught undergraduate geology in the laboratory and the field. Before joining OSIRIS-REx, she worked at the editorial-production nexus of the journal *Science* and as a science writer with the U.S. Global Change Research Program, where her team received an Award for Excellence from the White House for the 2014 National Climate Assessment.

Acknowledgments

The authors are grateful to the entire OSIRIS-REx team, past and present, for making the encounter with Bennu possible. For the images and data shown in this book, we especially acknowledge C. D. Adam, S. S. Balram-Knutson, O. S. Barnouin, K. J. Becker, T. L. Becker, E. B. Bierhaus, B. J. Bos, W. V. Boynton, K. N. Burke, S. R. Chesley, P. R. Christensen, B. E. Clark, M. G. Daly, D. N. DellaGiustina, C. Y. Drouet d'Aubigny, J. P. Emery, H. L. Enos, D. R Golish, V. E. Hamilton, K. Harshman, C. W. Hergenrother, E. R. Jawin, H. H. Kaplan, J. W. McMahon, M. C. Nolan, A. T. Polit, D. C. Reuter, B. Rizk, B. Rozitis, A. J. Ryan, D. J. Scheeres, J. A. Seabrook, A. A. Simon, and K. J. Walsh. We thank J. Tichá at the Klet Observatory for images of the Bennu discovery apparition and T. Gaither of the U.S. Geological Survey for assistance with feature nomenclature.

All images and data acquired by the OSIRIS-REx mission at Bennu are publicly available via the Planetary Data System at arcnav.psi.edu/urn:nasa:pds:context:investigation:mission.orex. Global digital terrain models and basemaps of Bennu can be explored by downloading the Small Body Mapping Tool at sbmt.jhuapl.edu.

NASA's Goddard Space Flight Center provides overall mission management, systems engineering, and the safety and mission assurance for OSIRIS-REx. Dante Lauretta of the University of Arizona, Tucson, is the principal investigator. The university leads the science team and the mission's science observation planning and data processing. Lockheed Martin Space in Littleton, Colorado, built the spacecraft and provides flight operations. Goddard and KinetX Aerospace are responsible for navigating the OSIRIS-REx spacecraft. Curation for OSIRIS-REx, including processing the sample when it arrives on Earth, will take place at NASA's Johnson Space Center in Houston. International partnerships on this mission include the OSIRIS-REx Laser Altimeter instrument from the Canadian Space Agency and asteroid sample science collaboration with CNES (France) and the Japan Aerospace Exploration Agency's Hayabusa2 mission. OSIRIS-REx is the third mission in NASA's New Frontiers Program, managed by NASA's Marshall Space Flight Center in Huntsville, Alabama, for the agency's Science Mission Directorate Washington.

Sir Brian would like to add his personal thanks to the following: Sally Avery Frost, Pete Malandrone, Phil Webb, Jamie Cooper, and David J. Eicher. And special thanks to Dr. Patrick Michel.